A Handbook of

Organic Chemistry Mechanisms

Peter Wepplo

Curved Arrow Press
Princeton, New Jersey

ISBN-10: 0-9779313-3-1
ISBN-13: 978-0-9779313-3-0

Address inquiries to P. Wepplo,
info@CurvedArrowPress.com
45 Wilton St., Princeton, NJ 08540
Printed in the United States of America
10 9 8 7 6 5 4 3 2 1

Table of Contents

Index

Index

Index

Index

F-N

Named Reagents & Reactions

Index

Index

Index

Index

T-Z

Functional Group Index
By Preparation

Preparation of	Starting Material	Chapter
Alcohol		
	alkene	5.2, 5.15-19, 12.10-11
	carbonyl compound	8.1-4, 8.12, 8.26, 9.1-2, 11.1-5, 11.8
	Substitution reaction	3.15, 3.17-21, 3.23-24, 3.26
Aldehyde		
	acetal	8.8
	alcohol	12.3, 12.7
	alkene	12.9
	alkyne	5.24
	diol	12.12
	ester	11.10
Alkane		
	alkene	13.3, 15.1-3
	ketone	11.12
Alkene		
	alcohol	4.13,14,16
	aldehyde	8.5-6
	alkenyl or aryl halide	13.1-2, 13.4
	alkyne	11.11
	amine	4.17-18
	electrocyclic rxn, Diels-Alder rxn	7.1-15, 7.17-18
	halide	4.1-12
Alkyne		
	dihalide	4.20-23
	Substitution rxn	3.10, 3.17
Amine		
	aldehyde	14.11
	amide	11.6-7, 11.9, 11.13
	carboxylic acid	15.4-5
	halide	3.7^1, 3.12^2
	nitro	14.13
Aromatic Compound, Substituted		
	benzyne reaction	14.13
	electrophile substitution	14.1-9
	nucleophilic substitution	14.11-12
Carboxylic Acid		
	alcohol	12.2
	aldehyde	6.3, 6.7, 12.4
	alkene	12.9, 12.11
	amide	8.24-25
	diketone	6.10
	ester	8.21-23
	ketone	9.13
	nitrile	8.28

Functional Group Index
By Preparation

Preparation of	Starting Material	Chapter
Carboxylic Acid Amide		
	carboxylic acid	8.17-19
	ester	8.27
	nitrile	8.29
Diene		
	halide	4.11-12
Diol		
	alkene	12.10-11
	epoxide	3.18^3
Enone or Unsaturated Ester		
	aldehyde	9.1, 9.3
	halide	4.10
Ester or Lactone		
	alcohol	8.13-16, 8.20, 8.31, 10.8
	alkene	5.3
	carboxylic acid	8.13-16, 8.20, 8.31, 10.8
	ester	9.5, 9.7, 9.9-11
	halide	3.4
	ketone	6.1, 6.2, 6.4-6
Ether or Epoxide		
	alkene	12.6
	halide	3.2, 3.13-14, 3.16, 3.28, 14.12
Halide		
	alcohol	3.5, 3.22, 3.25, 3.27, 10.2, 10.6-7
	alkane	16.1-4
	alkene	5.1, 5.4-14
	alkyne	5.20-21
	arene	14.3
	halide	3.1
	ketone	9.14
Haloalkene or Arene		
	alkyne	5.20-21
	arene	14.3, 14.5, 14.9, 14.14
	dihalide	4.20-23
Imine, Oxime, Hydrazone		
	aldehyde or ketone	8.9, 11.12
Ketal		
	ketone	8.7
Ketone		
	alcohol	8.11, 12.1, 12.5, 12.8
	alkene	12.9
	alkyne	5.22-23

Functional Group Index
By Preparation

Preparation of	Starting Material	Chapter
	arene	14.1, 14.8
	carboxylic acid	13.6, 14.1, 14.8
	diol	6.8-9, 12.12
	enone	9.15-16, 13.5
	ketone	9.6, 9.8, 9.14, 9.17
	nitrile	8.30
Nitrile		
	amide	10.1
	halide	3.8-9
	ketone	8.10
	nitrile	9.12
Organometallic		
	halide	8.1[4], 8.3[4], 13.1-6
Phenol		
	amine	14.14
	aryl ketone	6.11-12
	ether	3.15
α-Substituted Carbonyl Compound		
	aldehyde	9.1
	ester	3.11, 9.5, 9.7-8, 9.10
	ketone	3.19, 9.2-4, 9.6, 9.8, 9.15-17
	nitrile	9.12
Sulfur Containing		
	halide	3.3, 3.6
	alcohol	10.3

[1] Requires an additional reduction reaction
[2] Step one of a two step Gabriel amine synthesis. It requires an additional hydrolysis reaction.
[3] A diol will result if methanol is replaced with water.
[4] See *Notes.*, p. 22.

Functional Group Index
By Starting Material

Reaction of	Preparation of	Chapter
Alcohol		
	alcohol	3.18
	aldehyde	12.3, 12.7
	alkene	4.13-14, 4.16
	carboxylic acid	12.2
	ester	8.20, 10.8
	ether	$3.2^5, 3.14^5, 3.16$
	halide	3.5, 3.22, 3.25, 3.27, 10.2, 10.6-7
	ketal	8.7
	ketone	$6.12^6, 12.1, 12.5, 12.8$
	sulfonate	10.3
Aldehyde		
	alcohol	8.2-3, 8.12, 9.1-2, 11.1
	alkene	8.5-6
	amine	11.9
	carboxylic acid	6.3, 6.7, 12.2, 12.4
	enone	9.1, 9.3
	ester	8.12
	ketone	9.4
	phenol	6.11
Alkane		
	halide	16.1, 16.4
Alkene		
	alcohol	5.2, 5.16-19
	aldehyde	12.9
	alkane	15.1-3
	alkene	13.1-2
	alkyne	4.20-21
	carboxylic acid	12.9
	diol	12.10-11
	epoxide	12.6
	ester	5.3
	halide	5.1, 5.4-9, 5.12-15, 16.2, 16.3
	ketone	12.9
Alkyne		
	alcohol	3.17
	aldehyde	5.24
	alkyne	3.10, 3.17
	halide	5.20-21
	ketone	5.22-23
Amine		
	alkene	4.17-18
	amine	9.4, 9.15, 11.9
	carboxylic acid amide	8.27

Functional Group Index
By Starting Material

Functional Group Index
By Starting Material

Functional Group Index
By Starting Material

Reaction of	Preparation of	Chapter
Nitrile		
	amine	8.28-29
	carboxylic acid	8.28-29
	carboxylic acid amide	8.29
	ketone	8.30
	nitrile	9.12
Organometallic		
	alcohol	8.1-4, 8.26
	alkane	13.3
	alkene	13.1-2, 13.4
	ketone	13.5-6
Phenol		
	ether	3.13
α-Substituted Carbonyl Compound		
	alcohol	3.19, 8.12, 9.1-2
	aldehyde	9.1
	alkane	9.8-10, 9.17
	amine	9.4, 9.15
	carboxylic acid	9.13
	enone	9.1, 9.3
	ester	3.11, 8.12, 9.5, 9.7-11, 9.16
	halide	9.14
	ketone	3.19, 9.2-10, 9.14-17, 13.5
	nitrile	9.12
Sulfonate		
	nitrile	3.8
	sulfide	3.6

[5] An alcohol must first be converted to an alkoxide, see Chapter 9.7, ethoxide propagation. Also prepared with sodium metal.
[6] An alcohol must first be converted to a peroxide with hydrogen peroxide and acid.
[7] A benzamide conversion to an aniline.
[8] A carboxylic acid must first be converted to an acid chloride or anhydride.

Preface

About the Book

This book is an adaptation of *A Guide to Organic Chemistry Mechanisms*. That book was a "how to" guide for learning organic chemistry reaction mechanisms. It was designed to maximize how our brains learn. It provided steps for learning. This book is more like a solutions manual of reaction mechanisms. It does contain some commentary, but the focus of this book is to provide the logical mechanistic details of organic reactions. It does not discuss the variations and context of the reactions that are generally provided in an introductory organic chemistry textbook. That content is extensive and contributes to the great length and depth of those books.

The reactions are grouped by reaction type because it is easier to learn a series of related reactions. If your book is not organized in that manner, then select similar examples from the table of contents or from the index. The index is broken into three parts. There is a typical alphabetical list of key words found in the book. There are two additional indices, one for functional groups by preparation and one by function group by reaction. If you wished to find methods for the preparation of an alcohol, go the functional group index listed by preparation. The left column list the functional groups to be prepared and the center column lists the starting materials for their preparation. On the right, are the experiment numbers listed by chapter. Similarly, if you wanted to react an alcohol, you could go the the functional group by reaction and find alcohol in the left column and the products in the center column and the experiment number listed by chapter on the right.

The format of *A Guide to Organic Chemistry Mechanisms* made it difficult to include commentary on the reactions. That commentary was confined to a Notes Appendix. In this book, I have been able to describe an alternate model for atomic structure that I have used to explain why reactions might take place as they do. While this model is logical and simple, I cannot guarantee that it will agree with the explanation your teacher may be seeking.

In comparing this book with other organic chemistry books, you will find a difference in focus. Many books may skip one or more steps in a reaction mechanism. They may also combine steps together with several sets of curved arrows. I have tried to include every step and to rationalize them. I have found it advantageous for some reactions to take the extreme that no two events can occur at the same time. This may preclude a reaction being concerted. For many reactions, that will make the regiochemistry, selectivity, or kinetics easier to understand. If you find your teacher or your textbook describes the same reaction as concerted, you should accept that any steps that occur faster than atom movement may be considered as concerted. In the majority of those reactions, you will find the additional step I have introduced contains a fast step in which a new bond forms only from electron movement. Overall, these reactions may look different, but will in fact be identical.

Writing Style (and Meaning)

I also wrote this book in the first person. Why write in the first person? First of all, it is a format that I am comfortable with. I like how it sounds. I also want to write in the first person because science can be gray. A scientific proof may not be as strong as we would like it to be. I think we are too frequently willing to accept *something as true simply because it is written in a book*. By writing in the first person, you will have a natural sense that an idea is my idea and other scientists may not accept it. As some of the mechanisms, topics and models contained in this text are different; I will leave it to you to determine whether they are useful or true.* I hope by doing so, you may go back to your regular textbook and measure the thoughts of that author in the same manner.†

The Curved Arrow

In this book, I use a simple modification of pushing electrons. Curved arrows always start with a pair of electrons. If a traditional curved arrow starts with a pair of electrons shared by two atoms, the curved arrow is ambiguous. To which of the atoms will the new bond be connected? This ambiguity is avoided by using a "pre-bond" or dashed line that indicates which atoms will be joined. When used with a curved arrow, it shows which electrons move and which bonds are made or broken. As result, pushing electrons is consistent and unequivocal.‡

Rules for Pushing Electrons

To Make a Bond or Increase the Bond Order	
Two Electrons	One Electron
A curved arrow must start with a pair of electrons on an atom or connecting two atoms and end between two atoms. It may point toward a pre-bond to indicate a new bond or to a single, or a double bond for a change in bond order. A bond will form to the common atom of the starting material and product.	*Two half-headed curved arrows must end between two atoms and point toward a pre-bond, single, or double bond to indicate a new bond or a change in bond order. Each atom donating an electron are part of the new bond.*
To Break a Bond or Reduce the Bond Order	
Two Electrons	One Electron
A curved arrow must start with a pair of electrons connecting two atoms. That bond will be broken or reduced in bond order. If a curved arrow ends between two different atoms, a new bond is formed to the common atom. If not, no new bond is formed.	*Two half-headed curved arrow must start with a pair of electrons connecting two atoms. That bond will be broken or reduced in bond order. The curved arrows may end on an atom or connect to new atoms to form new bonds..*

Another difference between carried over from *A Guide to Organic Chemistry Mechanisms* is some of the valence electrons are not shown for common atoms as oxygen, sulfur, or the halogens. I have been careful to connect atoms together with pre-bonds. Including all of the valence electrons and pre-bonds could obscure the connection that bonds have to

---

* What is the difference? Models are never true, but they can be useful. If a model were true, then it wouldn't be a model for something.

† Ideas presented in peer-reviewed journals will contain a reference to their source. Therefore, a reader understands that the idea belongs to the source. If authors accept the principles first laid out, then they may become commonly accepted. However, in the strictest use of logic, it does not become more true. It will remain only as true as the original proof or proofs.

‡ Many books use this convention, especially with Diels-Alder reactions. I have added the term 'pre-bond.

atoms and not to the electrons. Secondarily, I used a single pair of electrons as a visual clue to students that they should start their curved arrows with that pair of electrons. I had also encouraged students, especially initially, is to circle the electrons being moved.

Consequently, you should find the reaction mechanisms are consistent and easy to understand. Since the curved arrows are a major difference in reaction mechanisms, I offered to convert all of the arrows to their conventional counterparts. It was the opinion of students, much like yourselves, that stated they liked the mechanisms with pre-bonds because they were "easier to understand".

A Handbook of Organic Chemistry Mechanisms© Peter Wepplo, 2009

1 — Getting Ready for Reactions

About the Atom

Before we start with reaction mechanisms, I want to review some basic facts about atomic structure and the resultant effects on chemical properties.

A Model for Acid-Base Reactions

Below is a simple dissociation reaction. The equilibrium constant (K) represents the ratio of ionized to unionized molecules. A large K corresponds with a high acid concentration. Since the pK_a is comparable to the negative log of K, a large K corresponds with a small pK_a and a strong acid. A large pK_a corresponds with a weak acid. If HX is a strong acid, the HX bond is very weak and easily broken. The ease of this reaction is reflected in the pK_a.

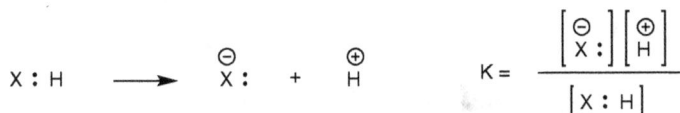

$$ X : H \longrightarrow \overset{\ominus}{X} : \quad + \quad \overset{\oplus}{H} \qquad K = \frac{\left[\overset{\ominus}{X}:\right]\left[\overset{\oplus}{H}\right]}{\left[X : H\right]} $$

pK_a Table

Compound	Bond length	pKa	Compound	Bond length	pKa	Halogen radius^2
H–C(–H)(–H)–H	1.10Å	50	H—Ï :	1.61Å	-10	1.33Å
H₂C=CH₂	1.08Å	44	H—Br:	1.41Å	-8	1.14Å
CH₂=CH–O–H	1.06Å	26	H—Cl:	1.27Å	-7	0.99Å
H—F :	0.92Å	3.2	H—F :	0.92Å	3.2	0.71Å

This table reveals a chemical paradox. In comparing HI, HBr, HCl, and HF, it is often explained that the longer the bond, the weaker it is and therefore the stronger the acid. However, the same comparison of the carbon compounds and HF (column one), shows the shortest bond (HF) is the strongest acid. *How can we resolve this paradox?*

While electron orbitals and hence bond lengths are more properly expressed by quantum mechanical calculations, I suggest that a simplistic model of Coulombic attraction between the positively charged nuclei and the electron pairs might also be used to predict a compound's bond length or strength, see *An Atomic Model for a Hydrogen Fluoride Bond*. Using Coulomb's Law, the bonding force(s) could be determined by the following equation.

$$F = \frac{kq_1q_2}{r^2} \quad \left(\begin{array}{l} \text{\textit{k is a constant, } } q_1 \text{ \textit{and} } q_2 \text{ \textit{are the}} \\ \text{\textit{charges, and r is the distance}} \end{array} \right)$$

In the ionization of HF, this model has two different forces at work, a fluorine–electron pair and a proton–electron pair force. For HF, $q_1=1$, $q_2=2$ for **proton**-electron pair affinity, and $q_1=9$, $q_2=2$ for **fluorine**-electron pair affinity (disregarding repulsive forces of other electrons), but the distance is not known. *Which force must be greater?*

An Atomic Model for a Hydrogen Fluoride Bond (HF).

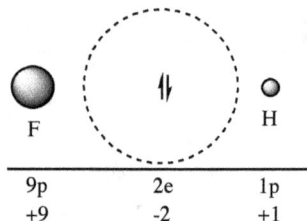

9p	2e	1p
+9	-2	+1

Since we know HF ionizes into H^+ and F^-, we know the electrons remain with the fluorine and therefore, the fluorine-electron pair force is greater. We can make a similar argument for HI, HBr, HCl, and HF. Because HI is a stronger acid than HF, we can conclude that the proton-electron pair force is less for HI and therefore the proton-electron pair distance must be greater. The table also shows the radius of the halogen covalent bond. This is the distance at which the valence electrons of covalently bound halogens are believed to exist. The *difference* between that distance and the bond length can be a crude measure of the proton-electron pair distance in the halo-acids. That distance increases from HF to HI (0.21, 0.28, 0.27, and 0.28Å). While this not a rigorous proof, the distances do agree with the principle.

If the atomic model used for the hydrogen fluoride bond is applied to the carbon compounds in column one of the table, then because ethyne is a stronger acid than ethene or methane, the **proton**–electron pair distance must be larger for ethyne (HC≡CH) than ethene (CH_2-CH_2) or methane (CH_4). Two consequences are observed from this model. One, a decrease in the **carbon**-electron pair distance can increase the **proton**-electron pair distance and decrease the C-H bond length. Two, as the bond lengths become shorter, the compounds become stronger acids. You will find many books note an equivalent argument that sp-hybridized carbons are more electronegative than sp^2 or sp^3-hybridized carbons.

Returning to the original observation about the bond lengths versus acidity, it was not the bond lengths that mattered. An increase in proton-electron pair distance will increase acidity. When the atoms are in the same row, then the bond lengths decrease as the electrons

---

* This is only a model to show the factors used to calculate the attractive forces using Coulomb's Law. It is **not** a model of atomic structure.

are being pulled away from a proton. Therefore, the shortest bonds corresponded with the strongest acid. However, for the halo-acids, because each halogen has an additional completed electron shell, the outer valence electrons are pushed out compared to the prior less acidic halo-acid. This increase in the atomic radius masks a similar increase in the **proton**-electron pair distance. This increase in **proton**-electron pair distance is often overlooked as the reason for the increased acidity.

Carbanion and Carbocation Stability

The next issue I would like to address is carbanion and carbocation stability. We will see the stability can be predicted from the electron donating properties of hydrogen and carbon. Commonly, textbooks contain a statement similar to, "carbon is a better electron donor than hydrogen."

In the following table, the relative stabilities of several carbanions and carbocations are listed. If carbon is a better electron donor than hydrogen, then increasing the number of carbon substituents should stabilize a carbocation. This rational is consistent with carbocation stabilities. A tertiary carbocation is the most stable carbocation and methyl the least. Similarly, increasing the electron density on a carbanion would destabilize a carbanion. Therefore, a tertiary carbanion is the least stable and a methyl carbanion is the most stable.

Carbanion and Carbocation Stability

Carbanion Structure	Relative stability	Carbocation Structure	Relative stability
$H-\overset{\ominus}{\underset{H}{\overset{..}{C}}}-H$	1 (most)	$H-\overset{\oplus}{\underset{H}{C}}-H$	4
$H_3C-\overset{\ominus}{\underset{H}{\overset{..}{C}}}-H$	2	$H_3C-\overset{\oplus}{\underset{H}{C}}-H$	3
$H_3C-\overset{\ominus}{\underset{H}{\overset{..}{C}}}-CH_3$	3	$H_3C-\overset{\oplus}{\underset{H}{C}}-CH_3$	2
$H_3C-\overset{\ominus}{\underset{CH_3}{\overset{..}{C}}}-CH_3$	4	$H_3C-\overset{\oplus}{\underset{CH_3}{C}}-CH_3$	1 (most)

The Nuclear Charge Effect

The following table shows a simple pattern for the number of bonds and charges on an atom. You may also use it as a simple alternative to calculating formal charges. All cells in the table contain ten electrons, and each column has the same number of protons. For each column, the proton that was attached to the nucleus of the upper molecule is in the nucleus of the lower molecule. Because each column contains the same numbers of protons and electrons, the difference in the properties of the compounds in each column is a function of the electron differences.

Ten Electron Table

	Charge	protons	Charge	Bond length	protons	Charge	protons
	+1		0			-1	
C	H−C⊕H (H H above, H below)	11†	H−C−H (H above, H below)	1.10Å	10p	H−C−H⊖ (H above, H below)	9p
N	H−N⊕−H (H above, H below)	11p	H−N−H (H below)	1.01Å	10p	H−N−H⊖	9p
O	H−O⊕−H (H below)	11p	H−Ö−H	0.96Å	10p	H−Ö:⊖	9p
F	H−F⊕−H	11p	H−F̈:	0.92Å	10p	:F:⊖	9p

How are the electrons of CH$_4$, NH$_3$, H$_2$O, and HF different? Coulomb's Law specifies that the force for electrical charges is dependent on the size and sign of the charges and is inversely proportional to the distance. Therefore, it seems logical to expect that shifting the protons from around a nucleus to the interior should exert a greater attractive force and pull the electrons closer to the nucleus. Consequently, the bond lengths would be expected to decrease from CH$_4$ > NH$_3$ > H$_2$O > HF. Similar to the *pK$_a$ Table* discussed on page 1, the more the electrons are pulled in, the greater the acidity (pK$_a$ CH$_4$ 50, NH$_3$ 36, H$_2$O 16, and HF 3.2). So, the properties of the compounds differ by the degree to which the electrons are pulled into the nucleus.

This nuclear charge effect will apply for cations, anions, and neutral molecules. The addition or removal of a proton will alter the electron availability across the table. For example, the electrons of hydroxide are more available than water and in turn hydronium ion. Similarly, moving down a column decreases electron availability and the basicity decreases from CH$_3^-$ > NH$_2^-$ > HO$^-$ > F$^-$.

Neighboring Atom Effect

The pK$_a$ of trifluoromethane (CF$_3$H) is approximately 25 while that of methane (CH$_4$) is 50. Which proton-electron pair distance will be greater, methane or trifluoromethane? We can extend these principles to neighboring atoms. The pK$_a$ tells us the proton-electron pair distance must be greater in trifluoromethane than methane. We might expect this result if the carbon-fluorine electrons are pulled away from the carbon. Then the electrons of the carbon-hydrogen bond can be pulled in more tightly (similar to water and hydronium

---

† This is the less common carbocation obtained by protonation of methane. CH$_3^+$ is more common, but does not fit the principle of the table because it has lost a proton and two electrons.

ion). This decrease in **carbon**-electron pair distance can increase the **proton**-electron pair distance and increase the acidity.

If the fluorine atoms are replaced with methyl groups, the carbon-hydrogen distance is also altered in the predicted way. Since carbon is an electron donor, see Carbanion and Carbocation Stability page 3, the electron donation will increase the carbon-electron pair distance and decrease the carbon-hydrogen distance of isobutane [(CH₃)₃CH]. The tertiary C-H bond length of isobutane is longer. This effect is in agreement with the carbanion stabilities.

In the following example, which HOX compound would be most acidic, HOH (water), HOOH (hydrogen peroxide), or HOCl (hypochlorous acid)? This question asks what effect will the atom attached to the oxygen (H, OH, and Cl respectively) have on the other electrons attached to oxygen? If a pair of electrons on an oxygen atom were pulled away from the oxygen by another atom, then the remaining electrons would feel the attractive force of the oxygen more strongly (or the electron pair repulsion forces would be diminished). As a result, if the electrons are pulled away from the oxygen by another atom, the other electrons of the oxygen will be pulled in. If the proton-electron pair move toward the oxygen, this will increase the proton-electron pair distance and result in a stronger acid. Therefore, acidity will decrease in the following order, HOCl>HOOH>HOH.

If you had two nitrogen compounds with different base strengths (ability to attract a proton), then which nitrogen would hold its electrons more tightly? We know the electrons of nitrogen should be more basic than oxygen or fluorine because there are fewer protons in the nitrogen nucleus to hold them as close. Therefore, the base strength of an atom should correspond with the distance of the electrons to the nucleus. If two nitrogen atoms differ in their basicity, then the less basic nitrogen would have its electrons closer to the nucleus (like oxygen or fluorine).

Can electronic effects be passed through a neighboring carbon atom? In the following acid-base reaction, the C-H acidity is affected by the substituents on a neighboring carbon. The substituent effects are noted in the table. Thus, the electron withdrawing properties can be passed through several atoms, though with decreasing magnitude by distance.

Neighboring Atom Effect

 pK$_a$ 50 A	*No electron withdrawing effects are present. High pK$_a$*	 pK$_a$ 44 B	*Two hydrogens have been replaced by an sp²-carbon. It will withdraw electrons and increase proton acidity.*

|
pK$_a$ 17
C | *Carbon with two hydrogens have been replaced by an sp²-oxygen. Because oxygen is more electron withdrawing, it markedly increases the acidity.* |
pK$_a$ 22
D | *The neighboring hydrogen has been replaced by a more electron donating CH$_3$-group and the acidity decreases.* |

Although I haven't discussed resonance structures to this point, you will find that the principle of resonance has a great effect on atom acidity. The examples B, C, and D can also be discussed as resonance effects, see Resonance Structures.[‡] The result of resonance on a structure is that the electrons can be shared with a neighboring atom.

Alkene Stability

Why is 2-butene more stable than 1-butene? I suggest that electron donation can be used to explain alkene stability. We have learned that an sp^2-carbon is more electron withdrawing than an sp^3-carbon. We might anticipate that an electron donating sp^3-carbon should stabilize an sp^2-carbon. If we compare 1-butene to 2-butene, 1-butene only has one electron donating carbon while 2-butene has two. Thus, increased substitution should increase the stability of an alkene.

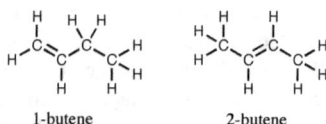

1-butene   2-butene

Atomic Charge

Formal charges are very effective in balancing the electrons in starting materials and products. However, there is an incorrect tendency to think a charge confers a property to the charged atom. In the *Ten Electron Table*, my objective was to compare structures with the same numbers of protons and electrons in each column. As a result, the properties of each structure was a function of how the electrons differed in each structure. It was easy to note how a nuclear-electron distance variable corresponded with the acidities of each compound. You can make the same observation about adding or subtracting a proton to a structure in the rows of compounds. Just as adding (or subtracting) a proton to a nucleus changed the Coulombic force, so will addition (or subtraction) of an external proton change the nuclear-electron distance and the formal charge.

What is hydrochloric acid? Dissolving hydrogen chloride in water gives hydrochloric acid, a solution of hydronium ions and chloride ions. Despite the opposite charges, this reaction with water shifts to the right and shows the electrons of water can form a stronger bond with a proton/hydrogen atom than a chloride ion can. The negative charge on the chloride ion tells us the electrons surrounding the chloride are in excess. However, because the electrons are in excess, we should not think that they should have a great affinity for cations (or that they are more basic than those of water). We should only expect the electrons of chloride ion are not held as tightly as the electrons of hydrogen chloride.

How does a formal charge apply to a structure? I have drawn the Lewis structures for hydronium ion, water, and hydroxide ion below. These structures all contain ten electrons and the oxygen nucleus has eight protons. The formal charges of the oxygen are positive, none, and negative while the true charge of the nucleus is unchanged for all three forms. The formal charges tell us about the electrons of oxygen. A proton (and other nuclei)

[‡] I sought a rational that should explain why the acidity should be increased and to not be an *ex post facto* resonance argument.

can change the nuclear-electron distance, but the charge of an electron and proton are unchanged. Therefore, protonation and deprotonation will decrease or increase the nuclear-electron distance, hence reactivity.

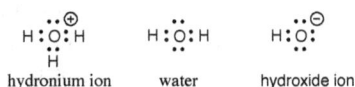

hydronium ion   water   hydroxide ion

What is hydronium ion? I have drawn an alternate formal charge form for hydronium ion. We know that hydrogen bonding is an important property of water. We should expect that if hydrogen bonding prevails in water, it will also be present with hydronium ions. These forms shift the formal charge from an oxygen atom to several atoms. This example shows some of the difficulty in representing molecules in a realistic yet compact form.

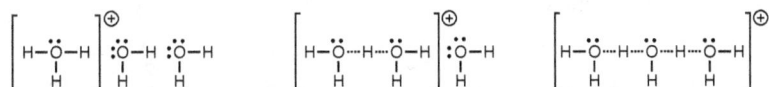

An error commonly associated with the representation of hydronium ion with a positively charged oxygen atom is to create a new bond to it. Despite a formal charge on an oxygen atom, it contains a completed octet and therefore cannot accept more electrons. Below is an example of an acid-base reaction proceeding to the right. Do not attempt to form a new bond to the oxygen atom.

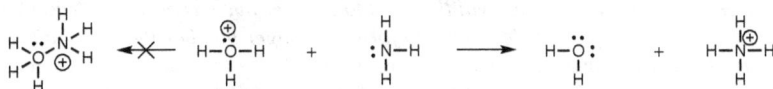

Ambident Anions and Amide Regiochemistry

The anion of acetone (see **D**, p. 5) , can be represented by two different forms, **D'** and **D"**. The pair of electrons are shared between the carbon and oxygen atoms. These forms can be interrelated by an electron shift and represent resonance structures, see further for a discussion of resonance structures. When resonance structures have a negative charge that

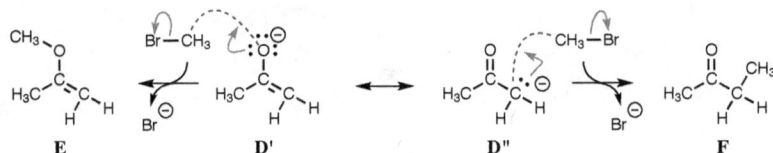

E   D'   D"   F

can be distributed over two different atoms, such as a carbon and an oxygen atom, they are referred to as ambident anions. The resonance structure with the negative charge on the more electron withdrawing atom will be the greater resonance contributor, **D'**. If this ambident anion were to react with bromomethane, we might expect **E** to be the product. However, the major product of this reaction will be **F**. Although **D'** may contribute more to the stability of this ambident anion, I reason that the electron pair distance will be greater on carbon than on oxygen. Therefore, a reaction of the (ambident) anions **D'** and **D"** will be more likely to occur on the carbon atom in **D"** because those electrons extend further from the nucleus. If a pair of electrons can exist on two different atoms[§], the less electron withdrawing atom (C>N>O) will be the more likely site of reaction.

That analysis works for most cases. However, a different result occurs on the reaction of a neutral amide (**G**) versus an amide anion (**H**), see further. I reason that the electrons of the amide nitrogen are depleted by the oxygen atom in the resonance structure **G'**. This electronic effect may be justified by features present in the IR and NMR spectra. The result is the non-bonded electrons of the nitrogen are held more tightly while increasing the oxygen-electron distance. Therefore, the reaction with an amide occurs on the oxygen atom.

If an amide anion is created by deprotonation of an N-H bond, the resonance structure **H'** shows that the oxygen is unable to deplete the non-bonded electrons from the nitrogen and the nitrogen remains the preferred reactive site. Predictably, amide anions are also more reactive than neutral amides.

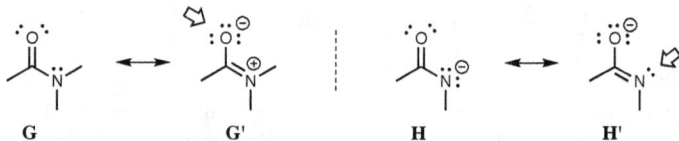

| G | G' | H | H' |

Summary about Atoms

A lesson I learned from how atoms make chemical bonds is '*the more tightly held a pair of electrons becomes, the weaker will the bond to another nucleus become*'. A corollary to that is '*a loosely held pair of electrons can form a stronger bond or the easier it becomes to form a bond*'. The analogy for an electron pair's ability to attack another nucleus is, '*chemical reactivity is like a boxer. A boxer with a longer reach would have an easier time to hit the nucleus of a another atom*'.

[§] This pattern generally works if the atoms of the ambident anion are in the same row (carbon, nitrogen and oxygen). However, if an oxygen were replaced with the more electronegative sulfur, the balance now favors reaction on sulfur. I presume that even though sulfur can donate electrons to another atom, because of the larger number of protons in its nucleus, it holds those electrons more tightly and reduces the electron density (and distance) on the other resonance atoms.

Resonance Structures

Drawing Resonance Structures

I have tried to simplify drawing resonance structures. I have included examples with simple repeating patterns for reference. It is important that everyone be able to draw the curved arrows and resonance structures, as they can be thought of as precursors to reaction mechanisms.

Anion Resonance Structures

The principle for understanding resonance structures is to understand that electrons will operate by a push-pull mode or model. If there is a net negative charge, start the curved arrow with those electrons and push toward a pi bond. Continue to move them toward (push) any neighboring pi bonds to create and break new bonds. You should note that two curved arrows are required to avoid structures with more than eight valence electrons.[¶]

For the following examples, curved arrows were added to show how the electrons move to form the next structure. For Examples 1-6, the first and last structures are the same. In that case, the curved arrows convert the intermediate resonance structure back to the starting structure. A circle was used for the non-bonded electrons. This was done to highlight their movement. You should think of a logical statement that is equivalent to the curved arrows. For Example 1, *"The non-bonded electrons on oxygen form a bond between the carbon and oxygen (now a double bond), and the electrons of the carbon-oxygen bond breaks and become attached to the oxygen."*

1.

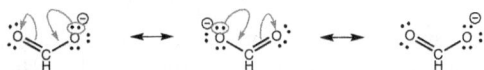

2.

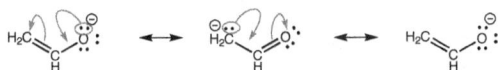

3.

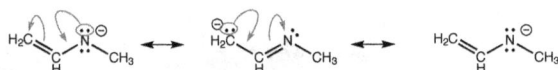

4.

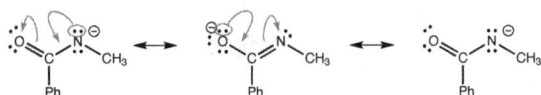

5.

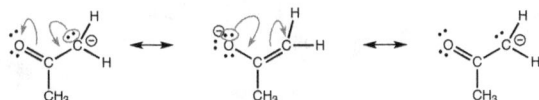

[¶] More generally, you should resist increasing the number of bonds or electrons on a receiving atom unless it contains a positive charge.

6.

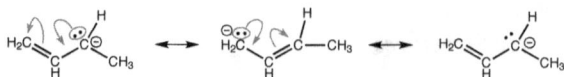

7.

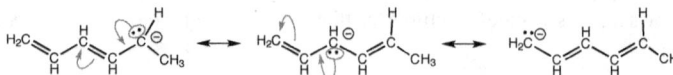

8.

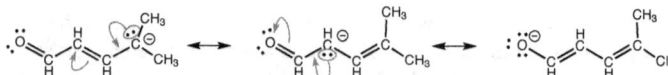

These examples show how the electrons can move, where the resulting charge will form, and how the charge can be distributed. It does not tell you on which atoms the greater charge density might exist nor upon which atom a reaction might next occur. You will note that since the original structure had a negative charge, the only charge that exists on any of the resulting resonance structures is a negative charge.

Cation Resonance Structures

For cation strucutes, there is a positive charge. It is the positive charge that will attract electrons (pull). Start a curved arrow from a neighboring pi or (pair of) non-bonded electrons and bring it toward the positive charge. You will note that only one curved arrow is necessary to create a new resonance structure. Because the original structure is a cation, completing its octet should not require further electron movement.

For Examples 9 and 10, the first and last structures are the same. In that case, the curved arrows convert the intermediate resonance structure back to the starting structure.

9.

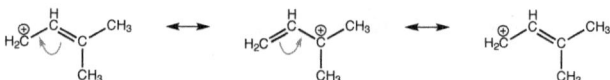

10.

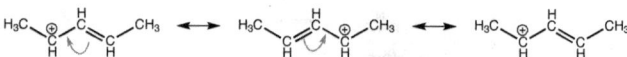

11.

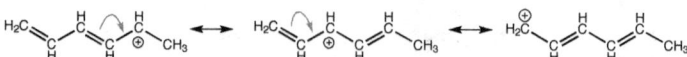

12.

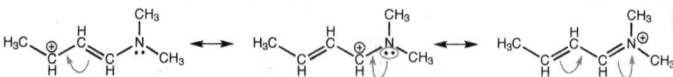

13.

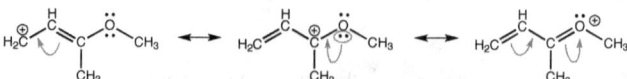

Again, these examples show how the electrons can move, where the resulting charge will form, and how the charge can be distributed. It does not tell you on which atoms the greater charge density might exist nor upon which atom a reaction might next occur. You will note that since the original structure had a positive charge, the only charges that exist on any of the resulting resonance structures are positive charges.

Neutral Resonance Structures, With Non-Bonded Electrons

If there isn't a charge but there are adjacent non-bonded electrons, then it will be the non-bonded electrons that will move (push) toward a neighboring pi bond. Start a curved arrow with the non-bonded electrons and direct them to the neighboring double bond. Do not exceed the octet rule.

For Examples 14-17, the first and last structures are the same. In that case, the curved arrows convert the intermediate resonance structure back to the starting structure.

14.

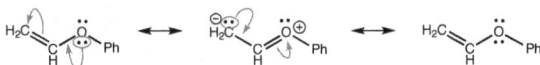

15.

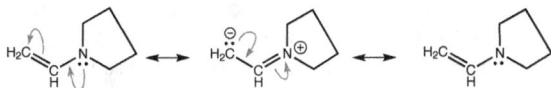

16.

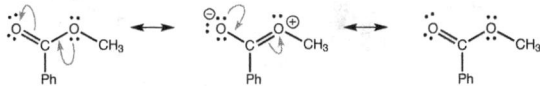

17.

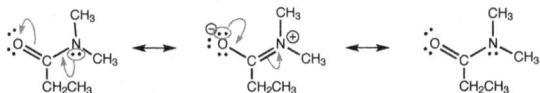

18.

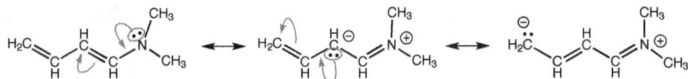

19.

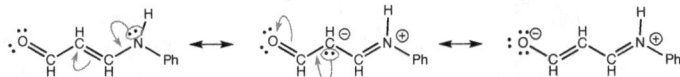

These examples show how the electrons can move, where the resulting charges will form, and how the charges can be distributed. It does not tell you on which atoms the greater charge densities might exist nor upon which atoms a reaction might next occur. You will note that since the original structure was neutral, the net charges that exist on any of the resulting resonance structures are also neutral and only two atoms have a charge.

Neutral Resonance Structures, Without Non-Bonded Electrons

If no charge exists and no neighboring non-bonded electrons are present, then use the electrons of a double bond (push). Start a curved arrow with the pi electrons of the most substituted double bond and direct them toward the least substituted atom of that double bond.

20.

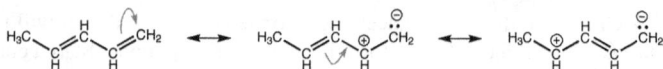

21.

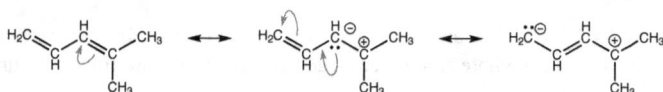

Example 22 is the same as Example 21 above, however the arrows are pointing in the opposite direction. If you are uncertain in which direction the electrons might move, a good strategy is to draw an arrow in the opposite direction and then to compare the results of the two possibilities. Compare the result below with the one above. Which is the more stable? Example 22 represents the lesser resonance contributor. It has a primary carbocation and a tertiary carbanion. If you do not recognize Example 22 as a lesser contributor, you may need to refer to Table *Carbanion and Carbocation Stability* (page 3) for carbanion and carbocation stability or to your textbook for the rules of resonance stability.

22.

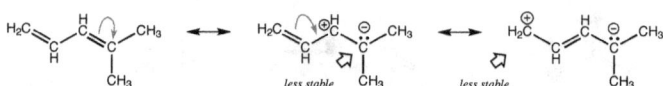

23.

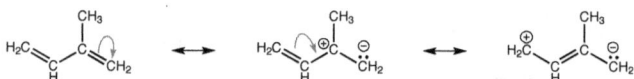

24.

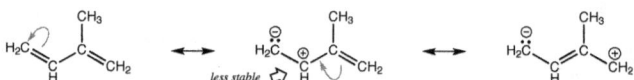

25.

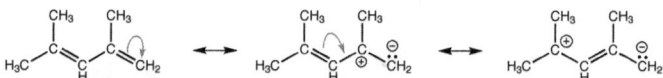

In these examples, we have incorporated chemical principles. The carbocations that are the most substituted are the most stable. Coinciding with this principle is that carbanions with the least substitution (or a heteroatom, last example) are the most stable.

Radical Resonance Structures

Radicals, compounds with unpaired electrons, are less stable than those with paired electrons. The fate of radical reactions is to form a paired-electron bond. However, sharing unpaired electrons with neighboring non-bonded or pi-bonds can attain added stability. Also note the curved arrow has a single barb indicating the movement of a single electron. Two arrows are required for a pair of electrons.

26.

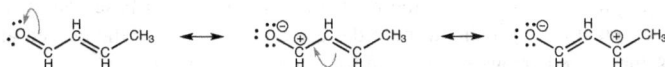

27.

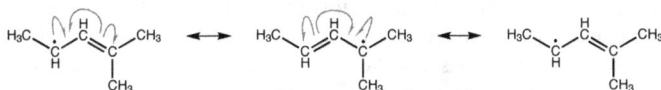

28.

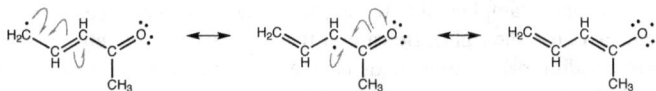

2 — Acid-Base Chemistry

Bronsted-Lowrey Acids and Bases

Acid-base reactions are often the first intermolecular reaction you will encounter. For Bronsted-Lowrey acids, the acid donates a proton and it exchanged from the strongest acid to the strongest base

In these examples, you must note the acid-conjugate base and base-conjugate acid that result in each reaction. You should note the use and meaning of the curved arrows. The meaning of the curved arrow is important for you to understand.

- The rule for predicting the product of an acid-base reaction is simple. A reaction will generally give the product that is the weakest base (or conjugate base). The base strength of a compound is related to the acidity of the acid, the stronger the acid, the weaker the base, or the corollary, the weaker the acid, the stronger the base. In order to compare the base strengths, the acidities of the acid and conjugate acid must be determined first.

- Look at each example and note the pK_a is under each acid (on the left) and conjugate acid (on the right). Be careful that you correctly identify the acid and the corresponding pK_a. Strong acids have a small pK_a and weak acids have a large pK_a. You may need to use a table to find some values.

- The molecule accepting the proton is the base or conjugate base. Compare the pK_a of the acid and conjugate acid. The strongest acid (smallest pK_a) corresponds with the weakest base. It is labeled, "weakest base". In Number 1, the pK_a of HF is 3.2 and the pK_a of acetic acid is 4.75. Because HF is the stronger acid (smallest pK_a), its conjugate base will be the weakest base. The equilibrium will shift to the right.

1. HF is the strongest acid as it has the lower pK_a. Therefore, F-, its conjugate base, is the weakest base.

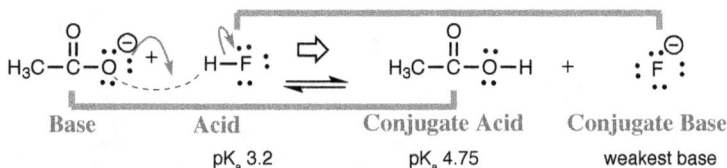

Base	Acid	Conjugate Acid	Conjugate Base
	pK$_a$ 3.2	pK$_a$ 4.75	weakest base

Notice the curved arrows. They describe the reaction that is taking place. We could write sentences to describe the bonds being formed and broken with the curved arrows.

A bond is being made between the oxygen and hydrogen atom with the electrons from the oxygen atom. A bond is being broken between the hydrogen and the fluorine atom with the electrons remaining attached to the fluorine atom.

I have written a number of additional examples. They show the bonds being formed and broken with the curved arrows and the direction of the equilibrium.

2. Label acids, bases, and conjugate acids and bases.

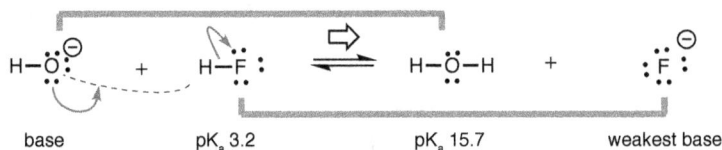

base pK_a 3.2 pK_a 15.7   weakest base

A bond is being made between the oxygen and hydrogen atom with the electrons from the oxygen atom. A bond is being broken between the hydrogen and the fluorine atom with the electrons remaining attached to the fluorine atom.

3.

weakest base   3.2   -1.7   base

A bond is formed between the oxygen and hydrogen with the electrons from the oxygen. The bond between the hydrogen and fluorine is broken with the electrons remaining attached to the fluorine.

4.

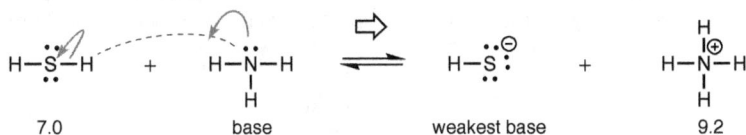

7.0   base   weakest base   9.2

A bond is formed between the nitrogen and hydrogen with the electrons from the nitrogen. The bond between the hydrogen and sulfur is broken with the electrons remaining attached to the sulfur.

5.

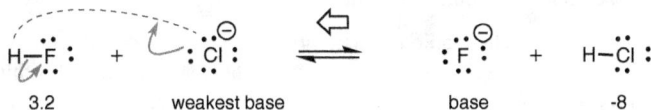

3.2   weakest base   base   -8

A bond is formed between the chloride and hydrogen with the electrons from the chloride. The bond between the hydrogen and fluorine is broken with the electrons remaining attached to the fluorine.

6.

10.0   base   weakest base   15.7

A bond is formed between oxygen-2 and hydrogen with the electrons from the oxygen-2. The bond between the hydrogen and oxygen-1 is broken with the electrons remaining attached to oxygen-1.

7.

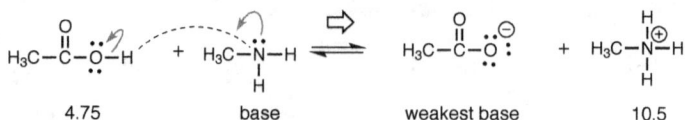

| 4.75 | base | weakest base | 10.5 |

A bond is formed between the nitrogen and hydrogen with the electrons from the nitrogen. The bond between the hydrogen and oxygen is broken with the electrons remaining attached to the oxygen.

8.

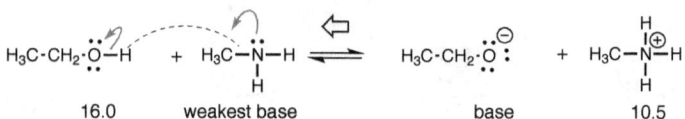

| 16.0 | weakest base | base | 10.5 |

A bond is formed between the nitrogen and hydrogen with the electrons from the nitrogen. The bond between the hydrogen and oxygen is broken with the electrons remaining attached to the oxygen.

9. For the purpose of predicting the direction of equilibria, I have left out the counter ions. In this example, an enolization takes place with butoxide. Similar reactions are present in Chapter 9 and a common source of butoxide is potassium t-butoxide.

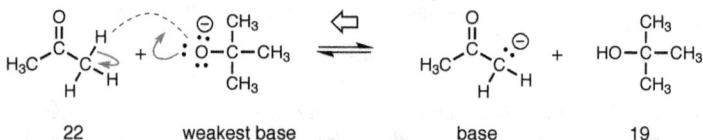

| 22 | weakest base | base | 19 |

A new bond is formed between the oxygen and hydrogen with the electrons from the oxygen. The bond between the hydrogen and carbon is broken with the electrons remaining attached to the carbon.

10. The sources for a methyl anion are commonly either methyllithium (CH_3Li) or methyl Grignard (CH_3MgBr). A hydrocarbon or ether solvent is used for this reaction because they are not more acidic than either of these reagents. It is also acceptable to use an excess of ammonia as the solvent. This is not the general method for the synthesis of sodium amide. However, other amide anions are made from amines and alkyllithium reagents.

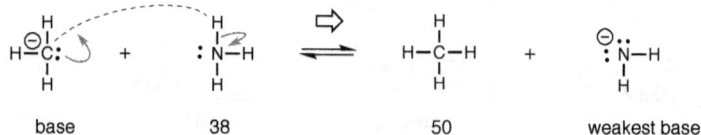

| base | 38 | 50 | weakest base |

A bond is formed between the carbon and hydrogen with the electrons from the carbon. The bond between the hydrogen and nitrogen is broken with the electrons remaining attached to the nitrogen.

11.

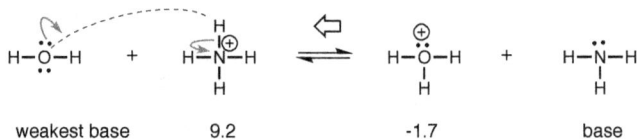

weakest base   9.2   -1.7   base

A bond is formed between the oxygen and hydrogen with the electrons from the oxygen. The bond between the hydrogen and nitrogen is broken with the electrons remaining attached to the nitrogen.

12.

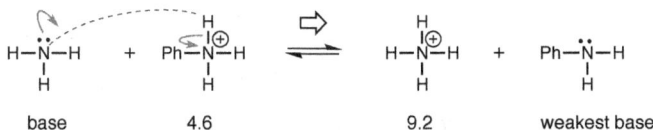

base   4.6   9.2   weakest base

A new bond is formed between the nitrogen-1 and hydrogen with the electrons from nitrogen-1. The bond between the hydrogen and nitrogen-2 is broken with the electrons remaining attached to nitrogen-2.

13.

4.75   base   weakest base   15.7

A bond is formed between the oxygen-2 and hydrogen with the electrons from oxygen-2. The bond between the hydrogen and oxygen-1 is broken with the electrons remaining attached to oxygen-1.

14. This is a good example of how an acetylide anion can be formed with methyl anion, usually from methyllithium (CH_3Li).

24   base   weakest base   50

A bond is formed between the carbon-2 and hydrogen with the electrons from carbon-2. The bond between the hydrogen and carbon-1 is broken with the electrons remaining attached to carbon-1.

15. This is step one of a Claisen condensation reaction, Chapter 9. You should note the equilibrium greatly favors the ester. Consequently, only low amounts of enolate are present in the Claisen condensation.

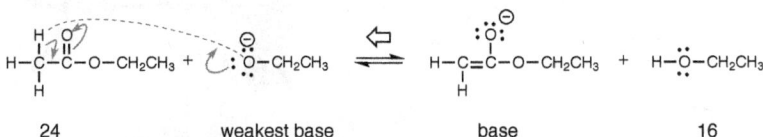

24   weakest base   base   16

16. This is the second acid-base reaction of a Claisen condensation reaction. You may wish to review this equilibrium when you study this reaction.

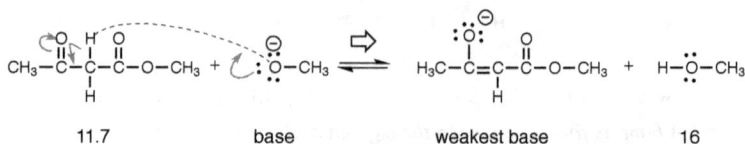

$$CH_3-\overset{O}{\underset{}{C}}-\overset{H}{\underset{H}{C}}-\overset{O}{\underset{}{C}}-O-CH_3 \ + \ \overset{\ominus}{:}\!O-CH_3 \ \rightleftharpoons \ H_3C-C\!=\!C-\overset{O}{\underset{}{C}}-O-CH_3 \ + \ H-\overset{..}{\underset{..}{O}}-CH_3$$

| 11.7 | base | weakest base | 16 |

17. Sometimes we may be unsure how a reaction might proceed. *Will H₂S and (CH₃)₂NH react together? What will the products be if they do?* In that case, draw ALL of the possible products and analyze the results. You will arive at the same final equilibrium even if different starting materials are used.

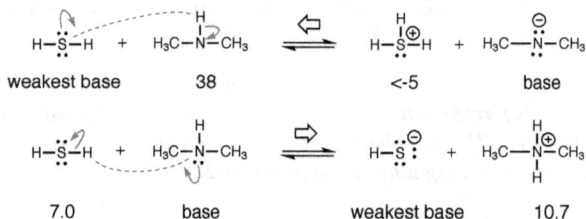

$$H-\overset{..}{\underset{..}{S}}-H \ + \ H_3C-\overset{H}{\underset{}{N}}\!\!=\!\!CH_3 \ \rightleftharpoons \ H-\overset{H}{\overset{\oplus}{\underset{}{S}}}-H \ + \ H_3C-\overset{..}{\underset{..}{N}}-CH_3$$

| weakest base | 38 | <-5 | base |

$$H-\overset{..}{\underset{..}{S}}-H \ + \ H_3C-\overset{H}{\underset{}{N}}-CH_3 \ \rightleftharpoons \ H-\overset{..}{\underset{..}{S}}\!\!:^{\ominus} \ + \ H_3C-\overset{H}{\overset{\oplus}{\underset{H}{N}}}-CH_3$$

| 7.0 | base | weakest base | 10.7 |

From the individual equilibria, you can combine them together to draw an overall result.

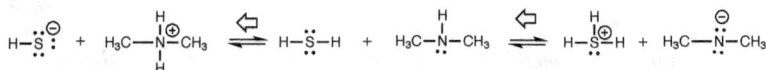

$$H-\overset{..}{\underset{..}{S}}\!\!:^{\ominus} \ + \ H_3C-\overset{H}{\overset{\oplus}{\underset{H}{N}}}-CH_3 \ \rightleftharpoons \ H-\overset{..}{\underset{..}{S}}-H \ + \ H_3C-\overset{H}{\underset{..}{N}}-CH_3 \ \rightleftharpoons \ H-\overset{H}{\overset{\oplus}{\underset{}{S}}}-H \ + \ H_3C-\overset{..}{\underset{..}{N}}-CH_3$$

18. *What is the equilibrium between HOCH₃ and CH₃NHCH₃?* First, determine the individual equilibria and then predict the overall result.

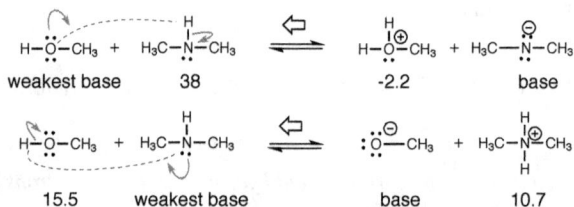

$$H-\overset{..}{\underset{..}{O}}-CH_3 \ + \ H_3C-\overset{H}{\underset{}{N}}\!\!=\!\!CH_3 \ \rightleftharpoons \ H-\overset{H}{\overset{\oplus}{\underset{}{O}}}-CH_3 \ + \ H_3C-\overset{..}{\underset{..}{N}}-CH_3$$

| weakest base | 38 | -2.2 | base |

$$H-\overset{..}{\underset{..}{O}}-CH_3 \ + \ H_3C-\overset{H}{\underset{..}{N}}-CH_3 \ \rightleftharpoons \ :\!\overset{..}{\underset{..}{O}}^{\ominus}-CH_3 \ + \ H_3C-\overset{H}{\overset{\oplus}{\underset{H}{N}}}-CH_3$$

| 15.5 | weakest base | base | 10.7 |

The equilibrium favors the neutral starting materials. However, the pK$_a$ differences are moderately separated between dimethyl ammonium cation and methanol. Therefore, dimethylamine will generate small amounts of methoxide.

$$:\!\overset{..}{\underset{..}{O}}^{\ominus}-CH_3 \ + \ H_3C-\overset{H}{\overset{\oplus}{\underset{H}{N}}}-CH_3 \ \rightleftharpoons \ H-\overset{..}{\underset{..}{O}}-CH_3 \ + \ H_3C-\overset{H}{\underset{..}{N}}-CH_3 \ \rightleftharpoons \ H-\overset{H}{\overset{\oplus}{\underset{}{O}}}-CH_3 \ + \ H_3C-\overset{..}{\underset{..}{N}}-CH_3$$

Lewis Acids and Bases

The prior exercise showed a hydrogen atom accepting electrons. With Lewis acids, other atoms can also accept electrons. See your text for further discussion.

19.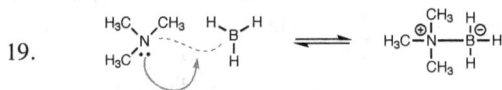

20.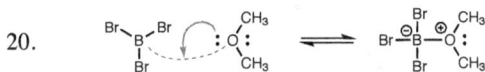

21.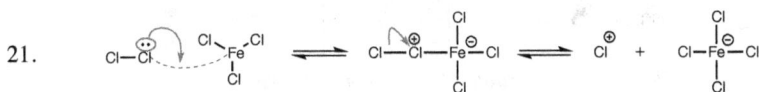

22.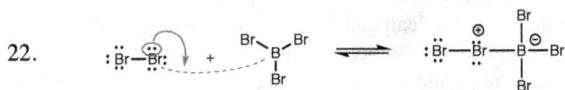

23.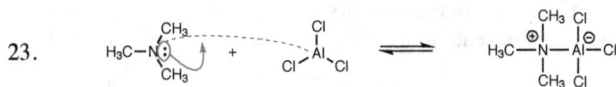

24.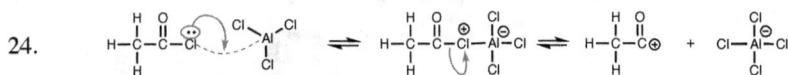

25.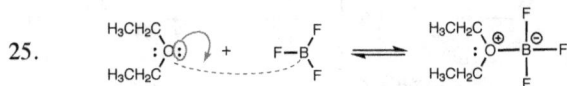

26.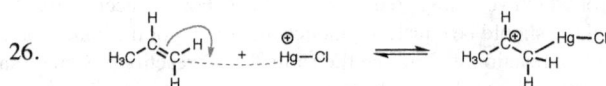

3 — Substitution Reactions

Substitution reactions can be described in two extremes, S_N2 and S_N1 reactions. An S_N2 reaction is like three billiard balls hitting together. The object ball does not move until the cue ball hits the stationary black ball first. There is an instant (at the transition state*) in which all three balls are in contact and then the leaving group leaves to form the product. In electronic terms, the electrons of the nucleophile form a bond with the carbon before the leaving group leaves. This is a single step reaction whose rate depends on the rate of collisions. Increasing the concentration or velocity of the reactants (heating) will increase the reaction rate.

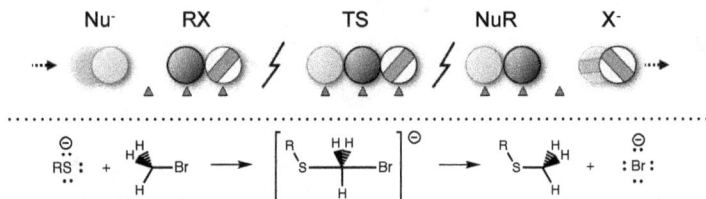

Nu, nucleophile; RX, substrate; TS, transition state; NuR, product; X, leaving group

An S_N1 reaction is like trying to find a parking space in a busy parking lot. You can't park until someone pulls out. In electronic terms, the leaving group leaves to form a carbocation (intermediate) before a nucleophile can add. This is the slowest step (the rate-limiting step). It is assumed that no bond formation occurs until a carbocation exists long enough for the leaving group to diffuse away and it does not influence any incoming nucleophile. If the RX compound had been optically active, then further reaction of the planar intermediate results in a product without optical activity. This is a two step reaction in which the rate of the reaction depends on the rate of the slow first step only, therefore the rate can only be changed its by changing the concentration of RX, solvent polarity, or temperature.

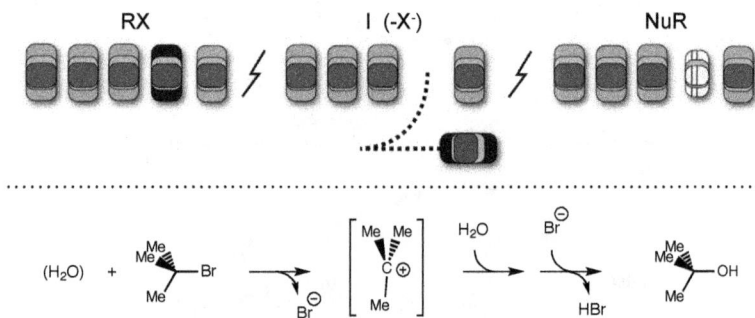

RX, substrate; I, intermediate; X, leaving group; NuR, product

These models represent the extremes of substitution reactions. In between these extremes will be the reactions for which you must predict the products. For a concerted reaction, the bond making and breaking should be equal. In practice, the timing of the bond making and breaking stages of an S_N2 reaction will not be the same for all reactions. Some transition states may resemble starting material or products more than others. This will depend on the

---

* A transistion state occurs at an energy peak in a reaction time profile while an intermediate is found in a valley.

electron withdrawing properties of the leaving group and the degree to which a nucleophile can donate electrons.

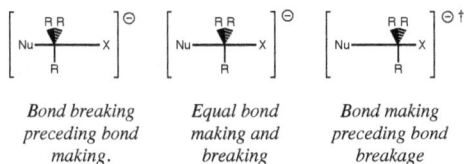

| Bond breaking preceding bond making. | Equal bond making and breaking | Bond making preceding bond breakage |

In S_N1 reactions, some inversion of a chirality center may take place. Therefore, participation by a nucleophile before the leaving group has completely diffused away may occur.

We can recognize some generalities. In an S_N2 reaction, the nucleophile initiates the reaction by pushing its electrons into the carbon nucleus of the RX group. The best nucleophiles are weak bases[‡], but with electrons that extend farther from the nucleus. Examples are iodide,[§] sulfide, thiocyanate, azide, phthalimide, cyanide, acetylide, malonate, and acetoacetate. The best leaving groups are iodide, bromide, sulfonate, or even neutral water. High nucleophile concentrations and polar aprotic solvents (DMF, DMSO) are common.

For the R-group, the ease of nucleophilic substitution is: methyl halide>primary halide>secondary halide. Groups that donate electrons to the carbon will slow an S_N2 reaction. Tertiary halides fail to react via S_N2 mechanisms. In order to displace a tertiary halide, the halide must leave before the nucleophile can attack (S_N1 reaction). If not, elimination will occur.

An S_N1 reaction requires conditions that favor carbocation formation, namely a good leaving group, a polar solvent, usually water, and weak nucleophiles (often uncharged). Displacements of tertiary halides are S_N1 reactions because they do not undergo S_N2 reactions. Similarly, primary halides (unless allylic or benzylic) do not undergo S_N1 reactions.

In between primary and tertiary are the secondary halides. They are the most ambiguous and the conditions of the reaction are most important in predicting the products. If the RX group is a secondary carbon, optically active, and the product retains optical activity but of the opposite configuration, then the reaction is an S_N2 reaction and an inversion process must have taken place. The presence of two carbons (electron donating groups, see Chapter 1) greatly slows the S_N2 reaction and only weakly basic nucleophiles give good yields of product. If a secondary halide is diluted, it will decrease the number of collisions necessary for an S_N2 reaction and can result in an S_N1 reaction.

For a given set of problems, students may not experience difficulties in substitution reactions. However, some may not write the product correctly because they are not completing the details of the bond making and breaking process. What follows are a series of substitution reactions. You should carefully note the bond making and breaking process that leads to the products.

---

† This model is not designed to be a true representation of the transition states. My objective is to show that an unequal bond making and breaking may occur. This can conform with the Hammond Postulate that transition states can be 'reactant-like' or 'product-like'. It also shades concurrent reactions in that some steps can occur before others.

‡ Elimination competes with substitution reactions and strong bases increase the elimination reaction, see Chapter 4.

§ These arguments contradict bond strength arguments of Chapter 1. However, the notion that the electrons of iodine and sulfur extend further from the nucleus does agree with their reactivity.

S$_N$2 Substitution Reactions

1. An S$_N$2 reaction of 1-chlorobutane with sodium iodide gives 1-iodobutane. This is a reversible reaction. However, it is usually run in acetone because NaI is soluble and NaCl is not. The NaCl precipitates and prevents the reverse reaction.

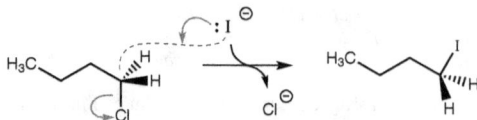

2. An S$_N$2 reaction of 1-chlorobutane with ethoxide gives 1-ethoxybutane (butyl ethyl ether). The displacement of a primary halide is subject to steric hindrance. As the nucleophile becomes larger, an increasing amount of elimination will accompany the substitution reaction, see Chapter 4.3.

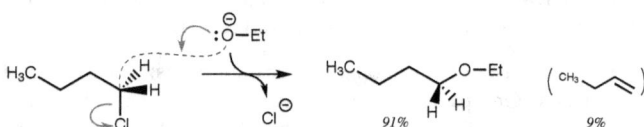

3. An S$_N$2 reaction of (R)-2-bromobutane with thiocyanate gives (S)-2-thiocyanatobutane. The example shows a variant of the principle discussed on *Ambient Anions*, p. 8. I rationalize this as sulfur, with its larger number of protons, as being able to deplete the electron density of nitrogen in the second resonance structure.

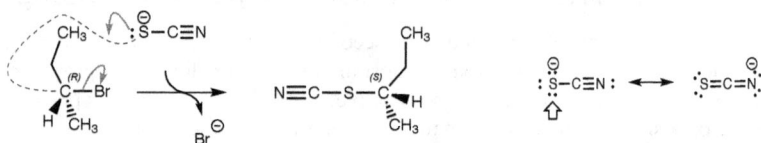

4. An S$_N$2 reaction of (S)-2-bromobutane with acetate gives *(R)-sec*-butyl acetate. If hydroxide or an alkoxide is used as the nucleophile, elimination would be the major product, see Chapter 4.2. Because the bromide is secondary, some elimination product is still found.

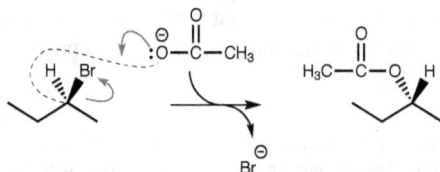

5. An S$_N$2 reaction of 1-butanol with hydrogen bromide gives 1-bromobutane. This example shows S$_N$2 reactions are not limited to basic conditions. If HCl is used, the reaction is not as efficient, therefore ZnCl$_2$ is used as a catalyst to accelerate the loss

of water. This comprises the Lucas test for primary, secondary, and tertiary alcohols. The mechanism will also change in going to a tertiary alcohol, see Chapter 3.22.

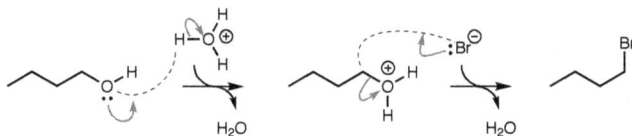

6.   An S$_N$2 reaction of a triflate with sodium methanethiolate gives a thioether with inversion. The electrons of the cyanide anion in Examples 6, 8, and 9 could react with the those of either the nitrogen or carbon. I argue that it isn't the negative charge that we write on the carbon that determines its reactivity, but rather the distance the electrons extend. As it has fewer nuclear protons, the reaction occurs on carbon. Also note that while the triflate can be displaced, the bromide cannot. The bromide is attached to an sp^2 carbon. *Direct substitution reactions cannot take place on sp^2 carbons*. Later we will see the mechanism for how a halide attached to sp^2 carbons can be displaced via an addition-elimination reaction.

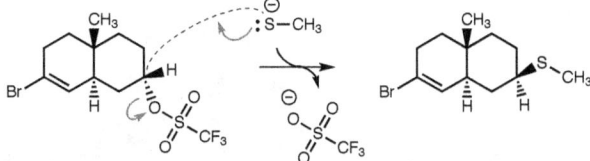

7.   An S$_N$2 reaction of *cis*-1-chloro-4-methylcyclohexane with azide gives *trans*-1-azido-4-methylcyclohexane.

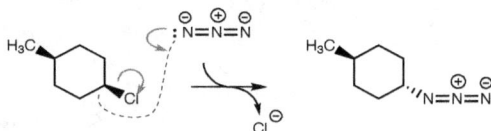

8.   An S$_N$2 reaction of a ditosylate with one equivalent of cyanide gives a mono-nitrile. Note, one equivalent signifies the mole to mole ratio. Axial substituents reacting faster than equatorial ones is a general trend. While a di-nitrile could be produced with excess cyanide, in this example, the axial tosylate reacts faster and leads to the major product.

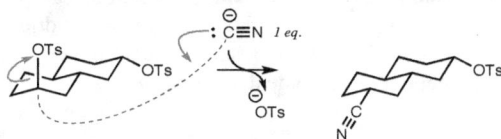

9.   An S$_N$2 reaction of 1-bromo-3-chloropropane with cyanide gives 4-chlorobutanenitrile. One equivalent signifies the mole to mole ratio. A bromide can react faster than a chloride and leads to the major product.

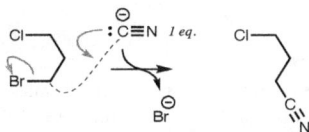

10. An S_N2 reaction of 1-bromobutane with acetylide anion gives 1-hexyne. The reaction of an acetylide anion is often the first anion used in carbon-carbon bond formation.

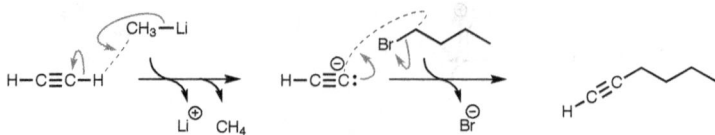

Methyllithium is a commonly used commercially available base for this reaction, see Chapter 2.14. However, depending on your class and textbook, you may find other bases are used. The same product can be obtained with sodium amide (left) or lithium diisopropylamide (LDA, right) for the same reaction. Sodium amide (sodamide) is an older base that is less frequently used in the modern organic chemistry laboratory than LDA.

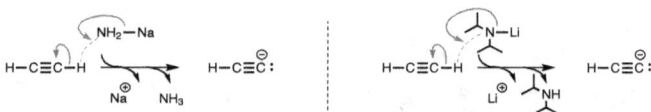

11. An S_N2 reaction of methyl phenylacetate enolate with iodopropane gives methyl 2-phenylpentanoate. The enolate of methyl phenylacetate is an ambident anion, see p. 8. Draw the resonance structure and answer, "Which electrons of the anion extend further from the nucleus, carbon or oxygen?" Also, note the order of the atoms change. All atoms must be legitimate Lewis structures.

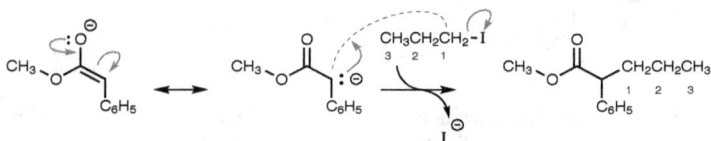

12. A Gabriel amine synthesis with formation of phthalimide anion (potassium carbonate) and alkylation with 1-bromo-2-butene. This is the first step (alkylation) of a Gabriel amine synthesis. You may find potassium hydroxide⌐ reported as base for a Gabriel amine alkylation reaction. I prefer potassium carbonate or sodium hydride as an alternate. Either can be used in situ. The greatest resonance contributor is shown. You should apply the principle reported for ambident anions, p. 8.

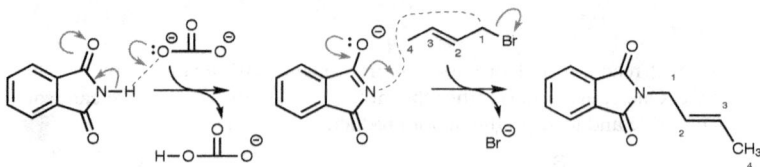

⌐ Anhydrous potassium hydroxide can be used. The intermediate potassium salt must be isolated in this procedure otherwise potassium hydroxide will hydrolyze the intermediate. The reaction I describe can be carried out in one flask without isolation of any intermediates.

Many references use a base hydrolysis for isolation of the amine. However, one of the best methods is a two step hydrolysis. Step one, base hydrolysis of the imide gives a carboxy amide and then step two, an acid hydrolysis of the amide, see Chapter 8.

13. An S_N2 reaction of bromomethane with phenoxide gives methoxybenzene (anisole). In this Williamson ether alkylation reaction, a carbon alkylation could take place as argued in ambient anions, *p. 8*, however, doing so would lose the energy gained from aromatization. The lower energy path preserves aromaticity by *O*-alkylation.

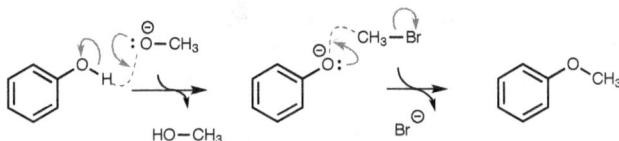

14. An S_N2 reaction of benzyl bromide with sodium ethoxide gives ethyl benzyl ether.

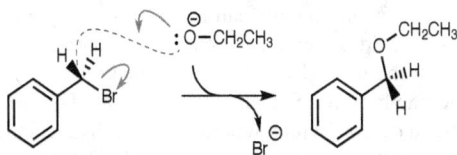

15. An S_N2 cleavage reaction of *p*-nitroanisole gives iodomethane and *p*-nitrophenol. The S_N2 reaction only cleaves the CH_3-O bond but not the aryl C-O bond. The aryl C-O bond is sp^2-hybridized and thus cannot undergo an S_N2 displacement reaction.

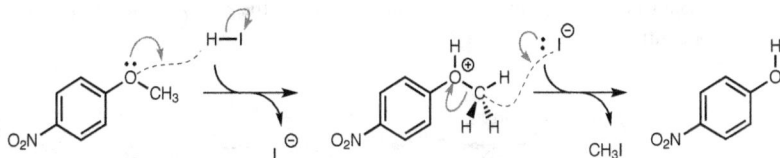

16. An S_N2 reaction of 1-bromo-2-butanol with sodium hydroxide gives 2-ethyloxirane (butylene oxide). Sodium hydride is a convenient and often used alternate reagent to form alkoxides. However, sodium hydroxide can be just as effective plus it is cheaper and safer to use.

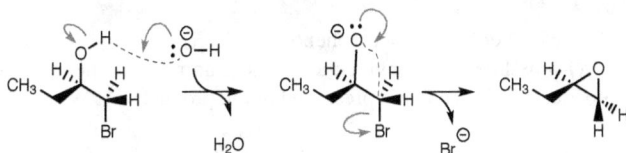

17. An S_N2 reaction of the anion of 1-butyne with *(R)*-2-ethyloxirane gives *(R)*-hept-5-yn-3-ol. The opening of the epoxide ring occurs at the least hindered carbon. I anticipate this S_N2 reaction will feature more bond formation preceding bond cleavage because there is less bond polarization of the carbon-oxygen bond and oxygen is not a good

leaving group. The reaction conditions (strong nucleophile) are consistent with an S_N2 reaction and sensitive to steric hindrance.

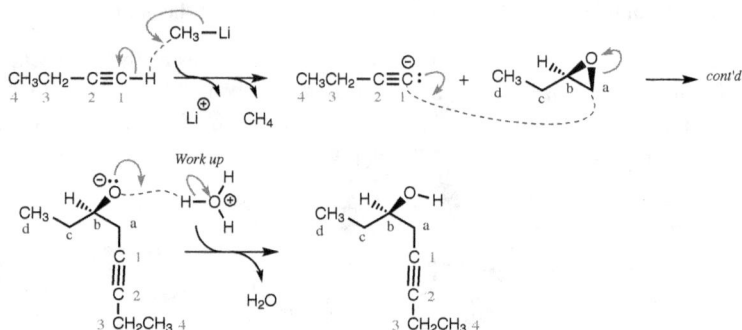

18. An acid catalyzed opening of 2-ethyloxirane with methanol gives 2-methoxybutan-1-ol. This is a good example of how an epoxide can react with a very weak nucleophile, an uncharged methanol. Epoxides aren't as reactive as primary halides so prior acidification will pull the electrons toward the oxygen to make it a better leaving group. This reaction blurs the line between S_N1 and S_N2 reactions. The reaction does not occur on the least substituted carbon, which would typify an S_N2 reaction, yet the reaction occurs with inversion of the carbon, which is not characteristic of an S_N1 reaction. The intramolecular nature of the oxygen atom leaving group allows the reaction dynamics to change as a result.

The reaction consumes and regenerates acid, since it isn't consumed only a catalytic amount is required, but the amount can affect the rate of reaction by increasing the concentrations of protonated species.

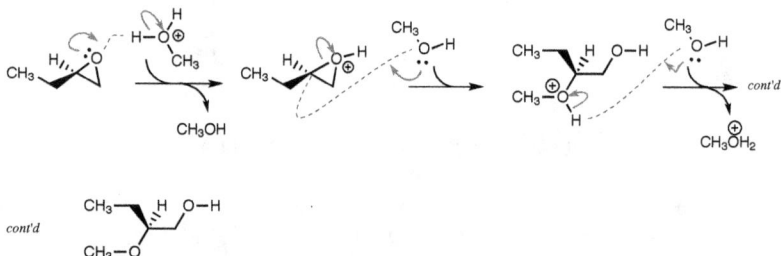

19. An S_N2 reaction of the enolate of acetophenone with (R)-2-methyloxirane (propylene oxide) gives (R)-4-hydroxy-1-phenylpentan-1-one is another another ambident anion. Note, the epoxide chirality center is preserved in the product.

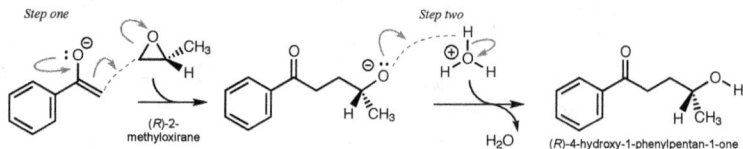

S$_N$1 Substitution Reactions

20. An S$_N$1 solvolysis reaction of *t*-butyl iodide gives *t*-butyl alcohol. A solvolysis reaction means the solvent causes the bond to break. In this example iodide is an excellent leaving group and gives a *t*-butyl carbocation intermediate. If iodide is replaced with poorer leaving groups, more alkene by-product will accompany the substitution reaction.

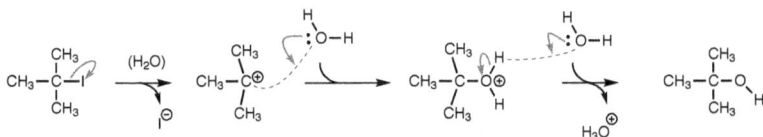

21. An S$_N$1 solvolysis reaction of *(R)*-(1-chloroethyl)benzene gives *rac*-1-phenylethanol.

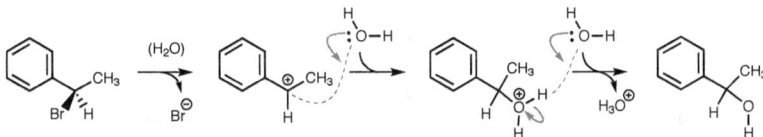

22. An S$_N$1 reaction of 1-methylcyclohexanol with hydrogen chloride gives 1-chloro-1-methylcyclohexane. Examples 20 and 22 show how reaction conditions control product formation. Both have tertiary carbocation intermediates yet different products.

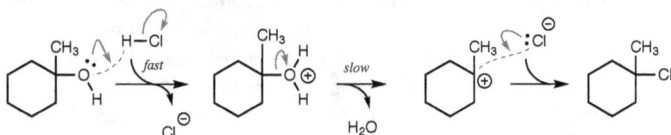

23. An S$_N$1 solvolysis reaction of *(1S,3R)*-1-bromo-1,3-dimethylcyclohexane gives *(1S,3R)*- and *(1R,3R)*-1,3-dimethylcyclohexanol. In this solvolysis reaction, the additional chirality center results in a *(3R)*-diastereomeric mixture.

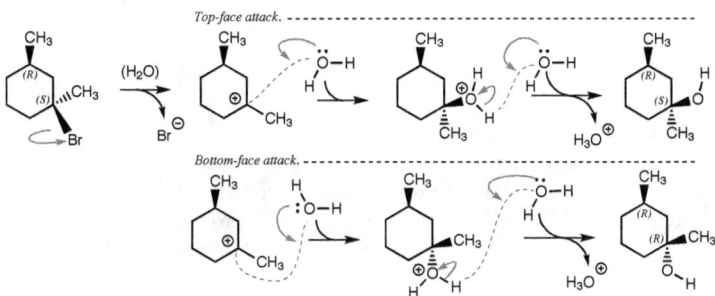

24. An S$_N$1 solvolysis reaction of 2-bromo-3-methylbutane gives 2-methyl-2-butanol. The products of S$_N$1 reactions can typically contain rearranged products. It can be difficult to predict the amount of rearrangement that might occur. In this example,

the rearranged intermediate is more stable than the unrearranged intermediate and is energetically favored. (*continued*)

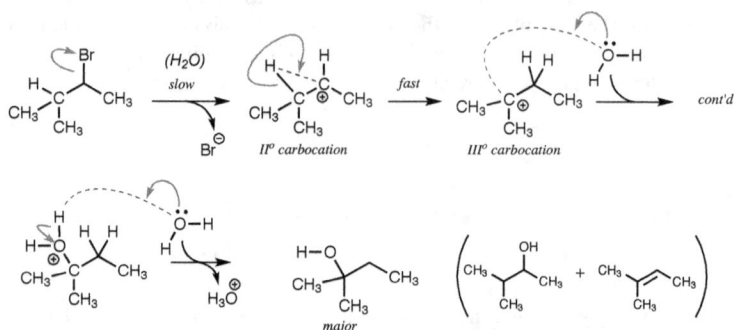

The kinetics of the reaction must be a slow solvolysis of the bromide consistent with an S_N1 reaction followed a fast rearrangement. The rearrangement must accompany bond cleavage or very soon thereafter as the reaction with water will also be a fast reaction. If the rearrangement were slow, then more unrearranged product would accompany the rearranged product.

25. An S_N1 reaction of 2-methylbut-3-en-2-ol with hydrogen bromide gives 1-bromo-3-methyl-2-butene. This example begins similar to Example 22, except the tertiary carbocation is also allylic. The bromide can attack at either carbocation of the resonance structures (not shown). The fastest reaction will occur at the more positive carbon. However, the C-Br bond will be weak with the substituents donating electrons to it. That product can reform the carbocation (as in Example 20). The competing slower reaction on the primary carbocation leads to the more stable (thermodynamic[**]) product, see *Alkene Stability*, p. 6.

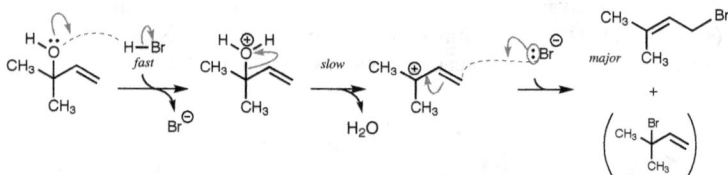

26. An S_N1 solvolysis reaction of 1-bromo-3-methyl-2-butene gives 2-methyl-3-buten-2-ol. This reaction appears to be a paradoxical result to the same intermediate of Example 25. However, where the conditions of Example 25 favored the thermodynamic product, this example 26 favors the kinetic product. Sodium carbonate is present in the reaction mixture, presumably to prevent protonation of the product alcohol

[**] See Chapter 5.10-11 for additional examples of thermodynamic and kinetic products.

and to reform the carbocation. Therefore, the product of this reaction is the kinetic product** of the reaction.

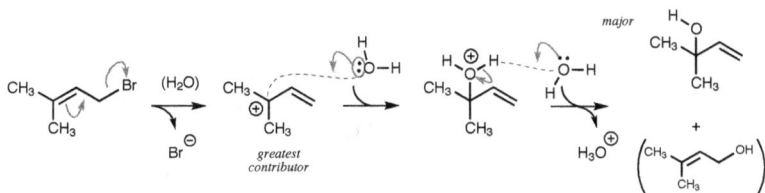

If the isomeric bromide were solvolyzed under the same conditions, the same products result. However. the tertiary-allylic bromide would solvolyze at a faster rate.

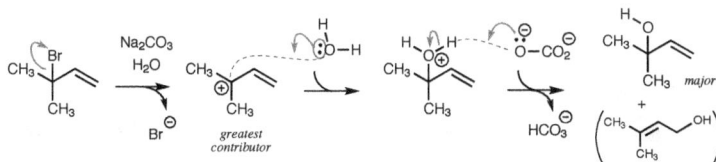

27. An S_N1 reaction of 2-methylcyclopentanol with hydrogen bromide gives 1-bromo-1-methylcyclopentane.

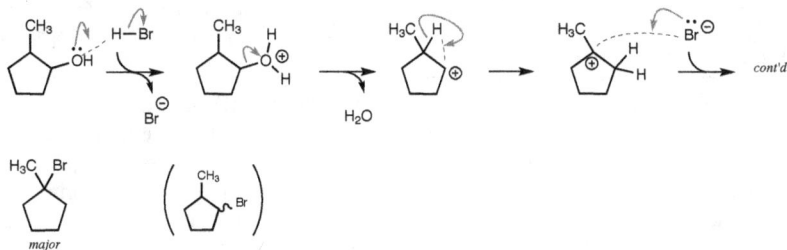

28. An S_N1 solvolysis reaction of a substituted 3-chlorocyclopentene gives a substituted 3-methoxycyclopentene. If the chloride were to give a symmetrical allylic carbocation, it could react with a nucleophile from either side and from the front or back. Therefore, if the chloride were optically active, a single racemic product would result. In this instance, a mixture of products will result.

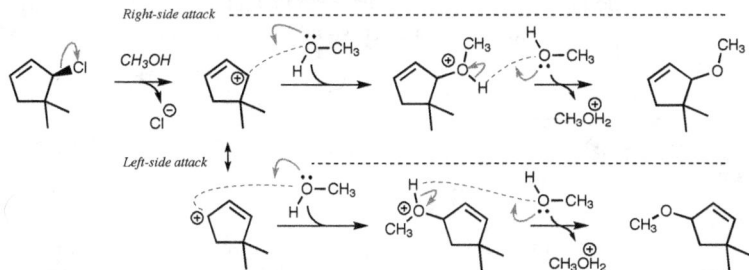

4 — Elimination Reactions

Elimination reactions have three variants[*], an E1cb, an E1, and an E2 elimination. In an E1cb elimination reaction, anion formation precedes the elimination step ($k_1 > k_2 = k_3$). In an E1 elimination reaction, carbocation formation precedes the elimination step ($k_3 > k_1 = k_2$). In an E2 elimination reaction, the elimination steps are concerted ($k_1 = k_2 = k_3$). In practice, we may find the rates can vary from those indicated.

Type	Example	Comments
E2		*Concerted reaction. The base and leaving group participate equally in the elimination reaction.*
E1		*Two step reaction. The rate of the reaction is solely dependent on the breaking of the bond to the leaving group. Possible rearrangement and loss of stereochemical control may occur.*
E1cb		*A two step reaction in which the anion does not result in a concerted loss of the leaving group. Proton acidity is an important factor in determining product distribution.*

An elimination reaction can be likened to an intramolecular substitution reaction. The electrons that were the nucleophile in a substitution reaction are now neighboring C-H electrons. A base is used to capture the proton while the electrons displace the leaving group in an S_N2-like reaction. The product is an alkene rather than a normal substitution product. Understanding these reactions also requires understanding the conformational constraints required for an intramolecular reaction.

Alkene Formation

1. An E2 elimination reaction of hydrogen chloride from 1-chlorooctadecane with potassium *t*-butoxide gives 1-octadecene. This example shows part of the challenge of predicting substitution and elimination reactions. With *t*-butoxide as the base, the elimination reaction is favored because of steric bulk and/or a decrease in solvent polarity. If the base were ethoxide, the product would be substitution, see Chapter 3.2.

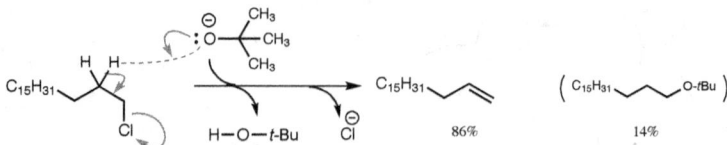

2. An E2 elimination reaction of hydrogen bromide from 2-bromobutane with sodium ethoxide gives *trans*-2-butene plus other butenes. In this example, several elimination factors are now combined. Since the halide is secondary, elimination and not substitution is the major product. Secondly, the stereochemistry of the elimination is

---

[*] You should consult a textbook for a more complete discussion of elimination reactions.

antiperiplanar (the proton and bromine are on opposite sides). Thirdly, according to the Curtin-Hammett principle, we might expect the product distribution to reflect the relative activation energy barriers for formation of the *cis*- and *trans*-2-butenes and not the ground state distributions.

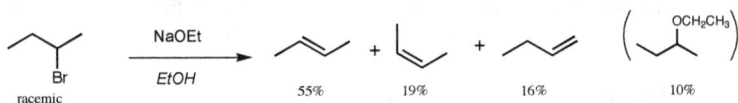

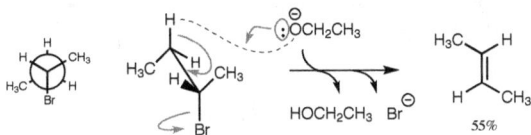

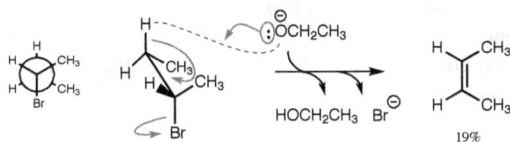

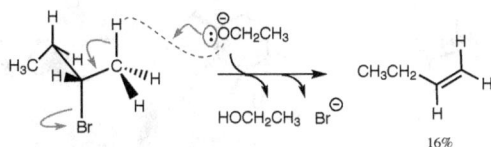

An alternate conformation leading to 1-butene might be considered as well. If the methyl group were *anti* to the bromide, this gauche conformation is only slightly higher in energy and an elimination to a 2-butene would be blocked.

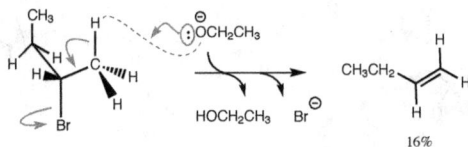

3. An E2 elimination reaction of 2-bromo-2-methylbutane gives 2-methyl-2-butene, a Zaitsev product. With a tertiary halide, the carbon-bromine bond is weaker,[†] especially in a polar solvent like ethanol. As a consequence, a greater positive charge will be present on the tertiary carbon to attract neighboring electrons. The electrons most readily donated are those attached to the most substituted carbon. Loss of the proton associated with the migrating electrons gives the product that is consistent with formation of the most substituted double bond. These conditions lead to a Zaitsev product.

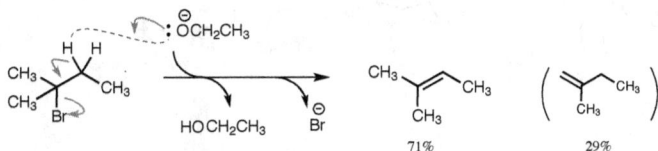

---

† The carbon-electron pair distance should be greater and result in a weaker bond, see Chapter 1.

4. An E2 elimination reaction of (*1R,2R*)- or (*1S,2S*)-1-bromo-1,2-diphenylpropane gives (*Z*)-1,2-diphenylpropene. This examples is consistent with the role that conformation has on the elimination and the regiochemistry of the product. A rotation must be made to get the hydrogen and bromide antiperiplanar. Elimination of HBr from that conformation leads to the *cis*-stilbene. A high concentration of reagents is necessary to ensure an E2 mechanism and the illustrated stereoselectivity.

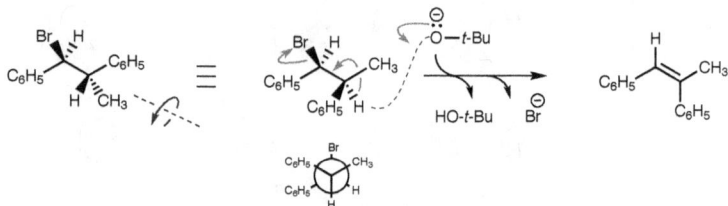

5. An E2 elimination reaction of *trans*-1-chloro-2-isopropylcyclohexane gives 3-isopropylcyclohexene. The conformation with an axial halide and hydrogen leads to the elimination product. That conformation places the isopropyl group axial and blocks elimination in that direction.

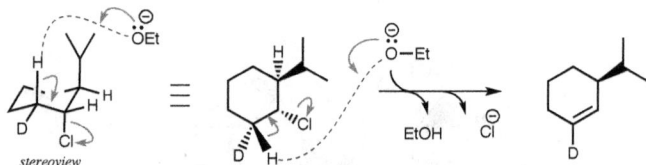

6. An E2 elimination reaction of *cis*-1-chloro-2-isopropylcyclohexane gives 3-isopropylcyclohexene. In this example, two axial hydrogens can give a mixture of the alkenes. The Zaitsev product is the major product.

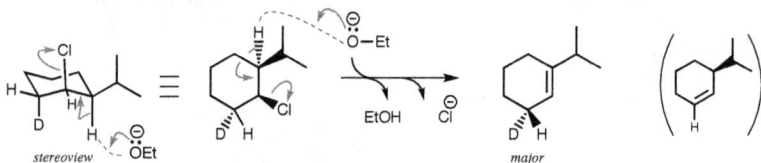

7. A competitive E2 elimination reaction of *cis*- and *trans*-1-bromo-4-*t*-butylcyclohexane and one equivalent of *t*-butoxide gives 4-*t*-butylcyclohexene and unreacted bromocyclohexane. The tertiary butyl group locks the conformation of the halide as axial and equatorial in each structure. The axial halide is eliminated 500x faster than the equatorial.

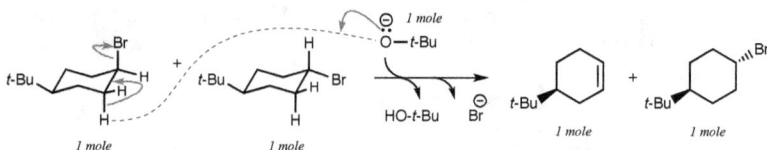

8. An E2 elimination reaction of 1-chloro-1-methylcyclohexane with ethoxide gives cyclohexene. These are Zaitsev elimination conditions and the more substituted alkene is the major product.

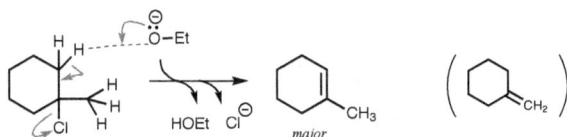

9. An E2 elimination reaction of 1-chloro-1-methylcyclohexane with *t*-butoxide gives methylenecyclohexane. This is the same chloride as Example 8, but a different product results with *t*-butoxide. The preferred elimination is from the least substituted carbon (most acidic, see Chapter 1).

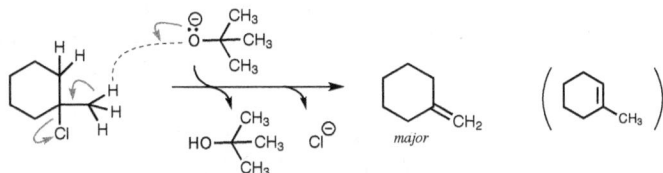

10. An E1cb elimination reaction of 3-chloro-3-methylcyclohexanone with *t*-butoxide gives 3-methyl-2-cyclohexenone. The hydrogen *alpha* to the carbonyl group is the most acidic and gives an enolate. Kinetically, this deprotonation is much faster than elimination. The rate of elimination depends on enolate concentration.

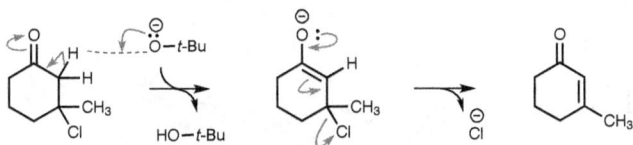

11. An E2 elimination reaction of 4-bromocyclohexene with *t*-butoxide gives 1,3-cyclohexadiene. A double bond is a variant of a carbonyl group and the increased acidity leads to the conjugated product.

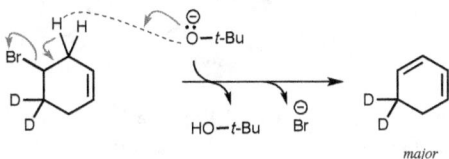

12. An E2 elimination reaction of 3-bromocyclohexene with *t*-butoxide gives 1,3-cyclo-hexadiene. This is a vinylogous* analog of Example 11. This elimination requires a shift of electrons (similar to resonance structures).

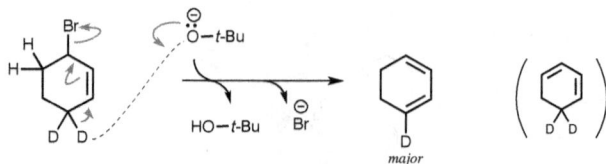

13. An E1 elimination reaction of 2-methylcyclopentanol by treatment with sulfuric acid gives methylcyclopentene. This is the first of several E1 eliminations. The first step is a fast protonation and then a slow, rate determining loss of water, to give a carboca-tion. Because water is such a weak base, the elimination step is slower than others. The requirement for an antiperiplanar deprotonation is lost. However, the deprotona-tion step does require a conformation in which the donated electrons can overlap with the carbocation.

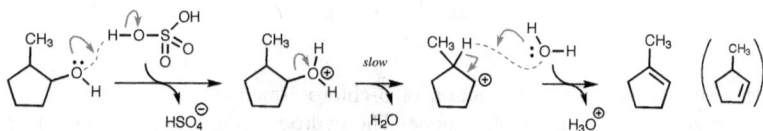

14. An E1 elimination reaction of 3-methyl-3-pentanol by treatment with acid gives a mixture of pentenes.

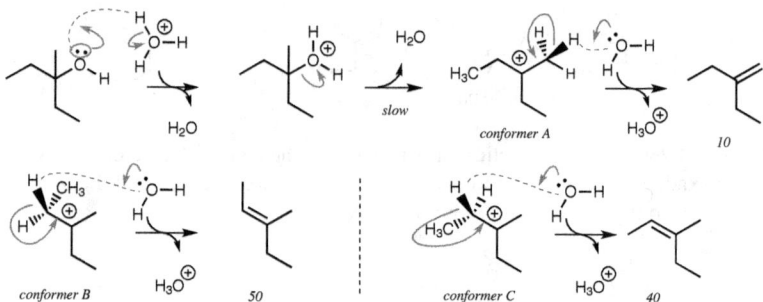

15. An E1 solvolysis-elimination reaction of a benzyl bromide in aqueous base gives methyl stilbenes. Since the carbocation is connected by a single bond, it is free to rotate and this results in a mixture of alkenes. This reaction should be compared to the E2 elimination in Example 5 from the same starting material. The solvents are usu-

---

* Vinylogous means the effect takes place *through* the double bond with the electrons of the double bond participating..

ally a water-organic mixture. If the hydroxide concentration is increased, the reaction could change into an E2 mechanism.

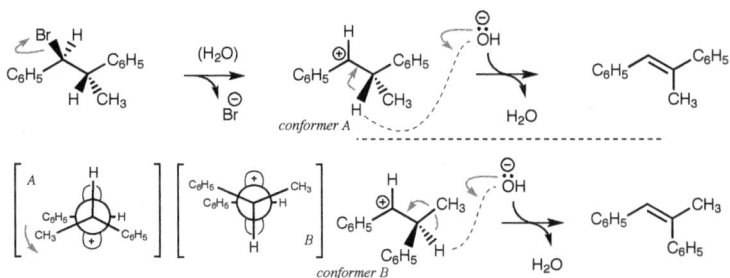

16. An E1 elimination reaction of 2-cyclobutyl-2-propanol and sulfuric acid gives 1,2-dimethylcyclopentene. The mechanistic steps start as shown in Example 13. Because a carbocation intermediate is present, one must be aware of possible rearrangement reactions. Rearrangement reactions that relieve ring strain are commonly used. The final deprotonation results in the most substituted alkene.

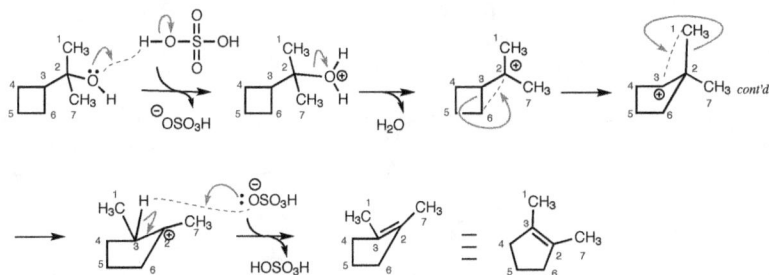

17. A Hofmann elimination reaction of a trimethylamine gives a 1-alkene. Reaction of N,N-dimethyl-2-pentanamine with iodomethane, silver oxide and elimination gives 1-pentene, the Hofmann elimination product. The difference in this example compared to the prior examples is the leaving group is a strong base, by comparison. The pK$_a$ of trimethylammonium ion is 9.76 while HCl is -7. I interpret this to mean the acidity of the proton being removed is a greater factor than that of a tertiary bromide, as Example 3. The product of the Hofmann reaction is the least substituted alkene.

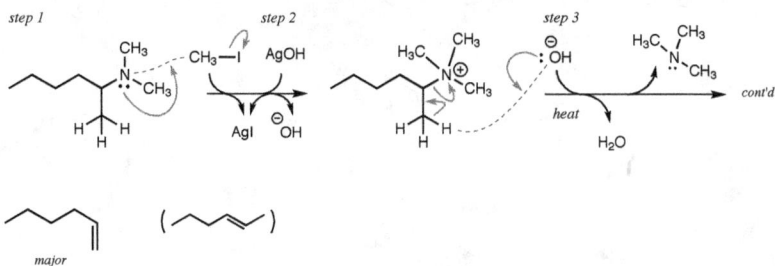

18. A Cope elimination reaction of a dimethylamine-N-oxide. Step 1, reaction of N,N-dimethyl-2,3-diphenylbutan-2-amine with hydrogen peroxide. Step 2, heating the N-oxide results in a cis-elimination reaction to give cis-α,β-dimethylstilbene, the

Zaitsev elimination product. Because the elimination uses a cyclic mechanism, a *cis*- or *syn*-elimination results. Acetate and xanthate pyrolysis, and Example 19 are also cyclic *cis*-eliminations.

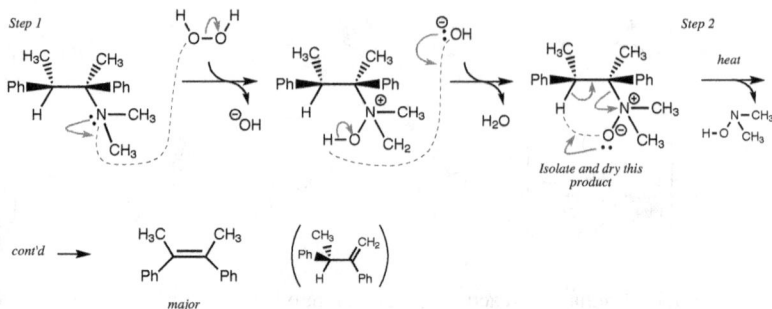

major

19. A selenoxide oxidation-elimination reaction of a cyclohexanone gives a cyclohex-enone via a *syn*-elimination. This is a good example of a selenium *syn*-elimination reaction. Selenium is readily oxidized and the selenoxide intermediate is readily eliminated at moderate temperatures. Sulfur will undergo an analogous reaction, but requires a higher temperature.

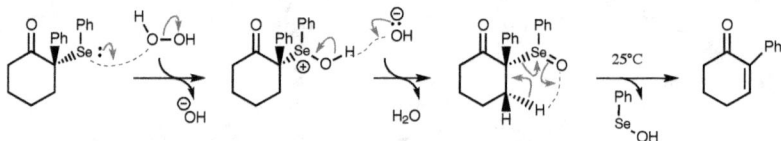

Acetylene Formation

20. A synthesis of 3-hexyne from *trans*-3-hexene by bromination and two elimination reactions. This sequence converts a double bond into a triple bond. The elimination is subject to the bromide configuration and conformation. The reaction could have started with geminal* dibromide, but because the configuration is frequently the result of a bromination reaction, that is the starting material shown. Go to Chapter 5.13 for the bromination mechanism.

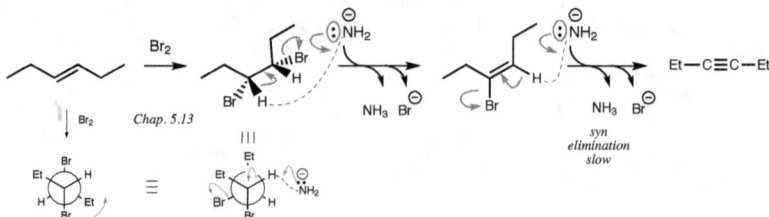

---

* Geminal means attached to the same atom.

21. A synthesis of 3-hexyne from *cis*-3-hexene by bromination and two elimination reactions. The *cis*-alkene give a Z-bromoalkene intermediate which results in a faster *trans*-elimination.

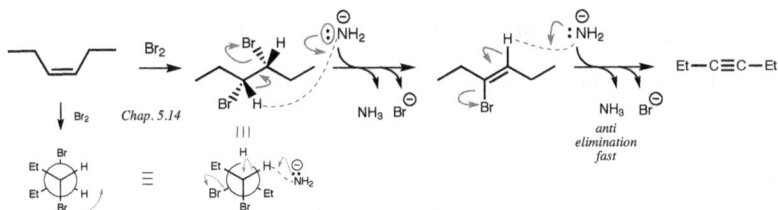

22. An E2 elimination reaction of 1,1-dibromopentane with *t*-butoxide gives 1-pentyne, a terminal acetylene. The elimination is achieved with *t*-butoxide, see Example 23. While it is not as strong a base, it also does not abstract the terminal acetylene hydrogen and therefore does not require an additional stoichiometric amount of base to be used.

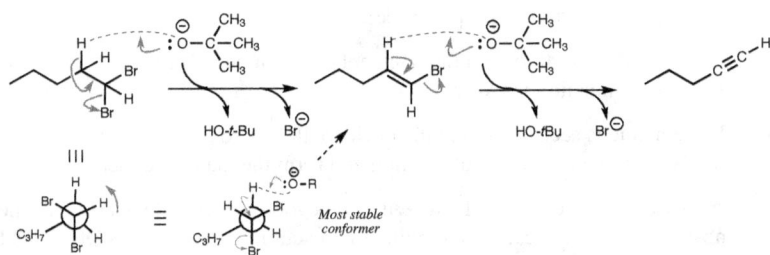

23. An E2 elimination reaction of 2,2-dibromopentane with LDA gives 1-pentyne. LDA is used as a convenient modern alternate to NaNH$_2$. In this reaction, because the base is strong and the acetylene is terminal, the product is the acetylide anion. By working up the reaction with D$_2$O, the expected terminal hydrogen is replaced with deuterium.

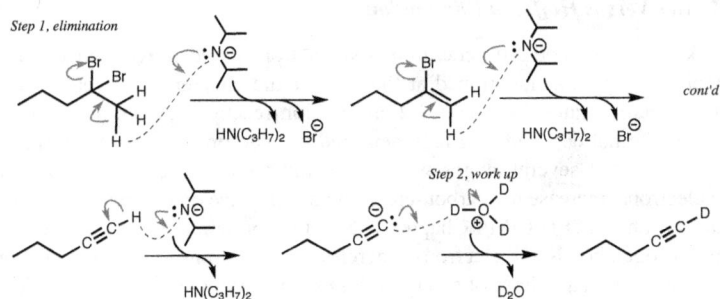

Elimination Notes

- Elimination reactions are similar to nucleophilic substitution reactions with iodide, bromide, or trifluoromethylsulfonate (conjugate bases of strong acids) as the best leaving groups.

- Increasing the temperature will increase the rate of an elimination.

- Increasing the solvent polarity, especially water, and decreasing the nucleophile concentration will increase the amount of E1 elimination, except primary aliphatic halides.

- The fastest reaction occurs when the electrons are donated from the opposite side of the leaving group. This is consistent with the antiperiplanar effects and *anti*-elimination reactions.

- A *cis*-elimination reaction is also common, though generally at a slower reaction rate. Examples 18, 19, 20, and 22 contain *cis*-elimination reactions.

Elimination versus Substitution

- If the substrate is a primary iodide or bromide, then an S_N2 nucleophilic substitution reaction is likely unless a sterically hindered nucleophile/base is used (tertiary butoxide, Example 1, or diisopropylamide).

- If the halide is tertiary and the nucleophile has a negative charge (except Cl^-, Br^-, or I^-), then elimination is likely.

- If the halide is secondary, and the nucleophile is strongly basic, then elimination will likely compete well with substitution and form the major product.

- A "hard" negative charge characteristic of a good nucleophile will also be most able to abstract a proton to cause an elimination reaction to occur, e.g, KOtBu and LDA.

- In a nucleophilic substitution reaction, elimination products can accompany the reaction. The best conditions for a nucleophilic substitution reaction are high concentration, polar aprotic solvents and a nucleophile with a negative charge. If the substrate is a primary halide, the substitution product is the likely product. If the halide is tertiary, then elimination is likely to occur. If the halide is secondary, then it may be more difficult to predict what the major product will be. Very basic nucleophiles will increase the amount of elimination products that will be found.

Zaitsev Versus Hoffman Elimination

Textbooks describe three distinct elimination reactions based on reaction kinetics, E1, E2, and E1cb. However, as mentioned at the start of the chapter, some reactions may vary from the specified rates, especially E2 elimination reactions. As an elimination reaction progresses, the halide bond will lengthen, and the carbon nucleus can attract electrons. I have represented several stages with numbered structures below. Carbon atoms can donate electrons, increase the carbon-electron pair distance of a carbon-halogen bond, and increase k_3. This was noted in Chapter 1. Substitution similarly affects the neighboring carbon-hydrogen bonds and electrons. Therefore the electrons most readily donated in an elimination reaction are those of a neighboring secondary or tertiary carbon. This will lead to the most substituted double bond or Zaitsev product. Furthermore, polar protic solvents are conducive to weakening the carbon-halogen bond and are often found with Zaitsev eliminations.

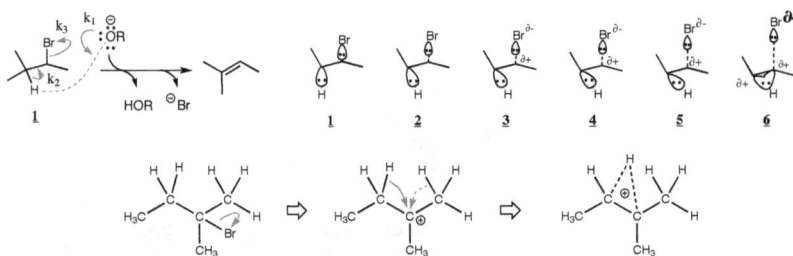

If the rate of k_3 is faster than k_1 or k_2, but not rate limiting, then an E2 elimination reaction will result and give a Zaitsev product. Examples 2, 3, and 8 represent conditions that lead to Zaitsev products. The halides are secondary or tertiary and the solvent is ethanol, all conditions that increase k_3.

If k_3 is slower than k_1 (or k_2), but still rate determining, then an E2 elimination reaction will result and give a Hoffman product. The less the carbon-halogen bond is stretched, the less the neighboring electrons (and k_3) will contribute to the transition state. Since that contribution increases the neighboring atom acidity (by going from **1** to **6**) and if no longer present, then increasingly k_1 will determine the product. Examples 9 and 17 represent this effect and lead to Hoffman elimination products.

Examples 8 and 9 are the same chloride and give different products. I believe solvent polarity may have a greater than expected effect in this result. As indicated, Example 8 gives the Zaitsev elimination product. This reaction was carried out in a polar protic solvent favoring an E1-like reaction. However, a mechanism can be changed by increasing the concentration of base to give an E2 elimination reaction. In Example 9, the solvent polarity was reduced by the change to t-butanol and k_3 should also be slowed. Consequently, hydrogen acidity (k_1) should have a greater effect on determining the product. The acidity of alkanes is primary>secondary>tertiary. Hoffman product formation agrees with hydrogen acidity. Similarly, Example 17 leads to a Hoffman product. In Example 11, a neighboring sp²-carbon increases k_1 to affect that product formation, though it is not a Hoffman elimination product.

One additional comment about Examples 9 and 17. The positive charge of the ammonium salt should not bias its electron withdrawing properties. Iodides and bromides give the least amount of Hoffman elimination products. From the table, as the conjugate acid (HX) of the leaving group (X⁻) becomes a weaker acid, the amount of Hoffman product increases. I believe this data agrees with the model suggested above.

RX	Zaitsev versus Hoffman	HX	pKa
RI	Zaitsev	HI	-10
RBr	Zaitsev	HBr	-8
RCl	Z or H	HCl	-7
RF	Hoffman	HF	3.2
$RNMe_3^+$	Hoffman	$HNMe_3^+$	9.8

5 — Electrophilic Addition to Alkenes and Alkynes

I have initiated several changes I have used in teaching organic chemistry into electrophilic addition reactions. Example 1 is a typical Markovnikov addition to an alkene. I believe that the traditional curved arrows do not teach students what the products should be. I found textbooks follow this reaction with discussions about cation stability to overcome the inherent ambiguity of the curved arrows. I have interpreted this discussion to mean that a student needed to know what the product of a Markovnikov addition was in order to understand the curved arrows. I believe pre-bonds (the dashed line that indicate where a new bond will form) make it easier to understand these reactions.

Kinetics and reaction timing has been part of the discussion of substitution and elimination reactions. This distinction is important in this chapter and is part of the discussion of individual reactions. While I expect some of you may disagree with my interpretation of these reactions, I want to ask you to skim through all of the reactions at some point and ask yourself these questions. *Should the alkene electrons react differently in reaction X than with reaction Y? If the electrons of an alkene or alkyne are the nucleophile, are the reagents also reacting as the electrophile?*

Addition of HX and H$_2$O to Alkenes

1. Addition of hydrogen bromide to propene gives 2-bromopropane.

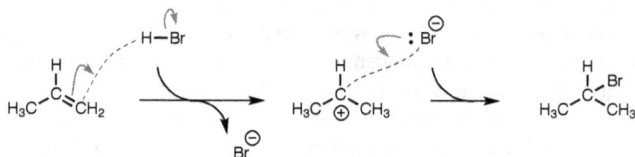

2. Acid catalyzed addition of water to methylcyclopentene gives 1-methylcyclopentanol.

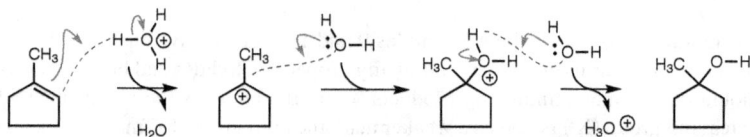

The hydration of a monosubstituted alkene is an example that may be found in some organic chemistry textbooks. Although this reaction can be preformed, it is not a general reaction. The alcohol is in equilibrium with the starting alkene and the alcohol is more basic . As a result, the reverse reaction is favored. If anhydrous sulfuric acid is used, a sulfate ester is formed. It is less basic and so blocks reverting to an alkene. However, the sulfate ester must be hydrolyzed and this would be an awkward reaction for common usage. In the laboratory, it would be easier to perform an oxymercuration-reduction (Example 16) or addition of acetic acid (Example 3) and hydrolyze the ester (Example 8.21 or 23).

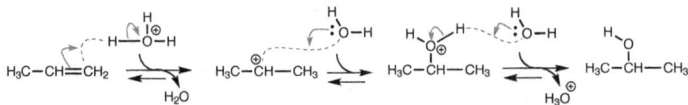

3. Addition of acetic acid to propene catalyzed by sulfuric acid gives 2-propyl ethanoate (isopropyl acetate). This is similar to Example 2, but because the alkene has fewer electron donating carbons, it is less reactive. Formation of an acetate ester as the product, is also less basic, and thus the reaction does not reverse.

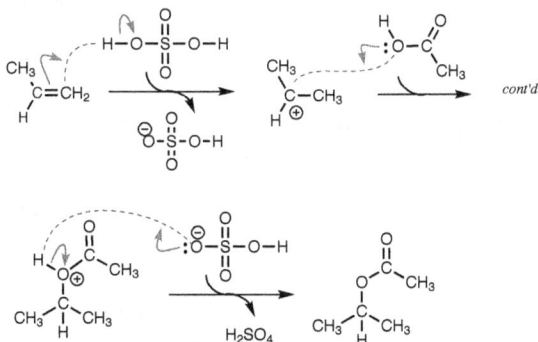

cont'd

4. Addition of hydrogen bromide to methylcyclohexene gives 1-bromo-1-methylcyclohexane. Electron donating carbons enables protonation of the alkene electrons. The intermediate tertiary carbocation is also stabilized by the electron donating carbons.

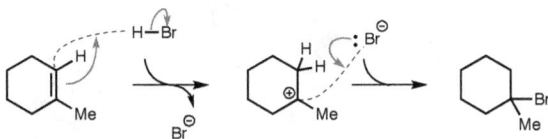

5. Addition of hydrogen chloride to (E)-3-hexene gives (R)- and (S)-3-chlorohexane. In this addition, a single chirality center is present. Chloride adds to either side of the planar carbocation so both R and S-isomers are formed equally.

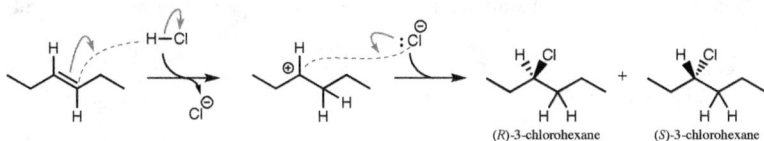

(R)-3-chlorohexane   (S)-3-chlorohexane

6. Addition of hydrogen chloride to (Z)-3-hexene gives (R)- and (S)-3-chlorohexane. The same products result from cis-3-hexene as from trans-3-hexene, see Example 5.

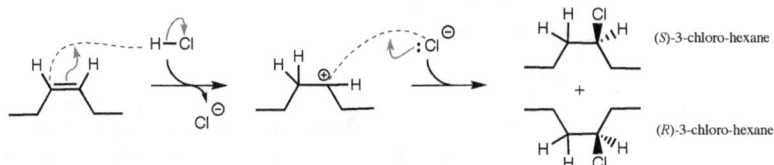

(S)-3-chloro-hexane

(R)-3-chloro-hexane

7. Addition of hydrogen bromide to 3-methyl-1-butene gives after rearrangement, 2-bromo-2-methylbutane. Often, I am uncertain whether a rearrangement should take place. This is a frequently used example. You can decide from the product mixture how difficult it can be to predict whether a rearrangement should take place.

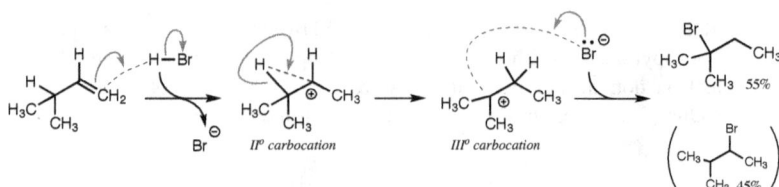

8. Addition of HBr to 2-cyclobutylpropene gives, after rearrangement, 1-bromo-1,2-di-methylcyclopentane. This example typifies a kind of rearrangement reaction I find in many texts and examples. Relief of ring strain is another impetus for rearrangement. I usually assume that if a carbocation can form adjacent to a strained ring, then a rearrangement will occur.[*] I always suggest to students to anticipate a rearrangement reaction with relief of ring strain.

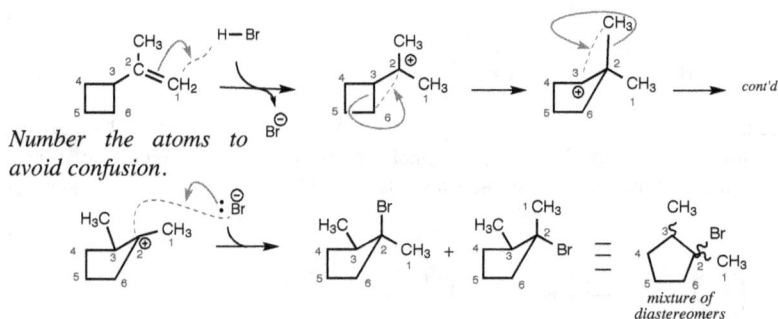

Number the atoms to avoid confusion.

9. Addition of hydrogen chloride to allylbenzene gives, after rearrangement, (1-chloro-propyl)benzene. This rearrangement probably occurs to the greatest extent compared to Examples 7 and 8. After rearrangement, the aromatic ring can donate electrons to stabilize an adjacent carbocation (you can draw resonance structures utilizing the aromatic ring). That stabilization gives greater impetus to a rearrangement.

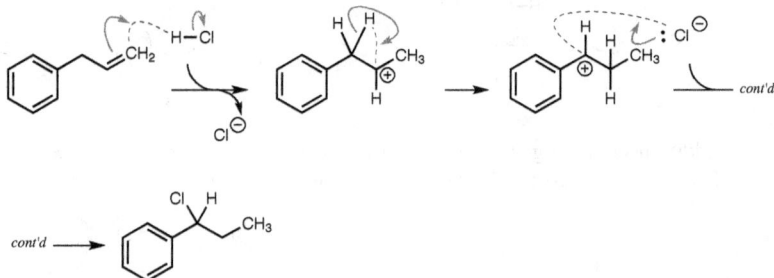

10. Addition of hydrogen chloride to 2-methyl-1,3-butadiene (isoprene) can give, 3-chloro-3-methyl-1-butene, the kinetic product, or 1-chloro-3-methyl-2-butene, the

---

* I do not have product distribution data. I would expect unrearranged products to also be present.

thermodynamic product. In most reactions, the kinetic and thermodynamic products are the same, but dienes are good examples for producing different thermodynamic and kinetic products. With dienes, a kinetic product can sometimes be isolated that can react further to give a more stable thermodynamic product.

If the activation energy of a reaction is low, the reaction will be fast. However, it may not be the most stable product. With dienes, that product may continue to react under those conditions to give a more stable product. How would you know that a kinetic or thermodynamic product should be result? A signal that a kinetic product might be expected is if the reaction is carried out at a low temperature.

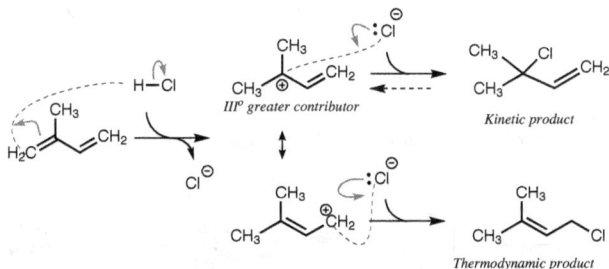

Kinetic product

Thermodynamic product

You shouldn't assume that an addition will occur in a 1,4-manner. In following example, the 1,2-adduct is the kinetic and thermodynamic product. You should assess the carbocation and alkene stability. The more stable the carbocation, the less stable will its product be. The more substituted the alkene, the more stable it will be. In Examples 10, the resonance structures contain a tertiary versus a primary carbocation. One might expect formation of a product at the tertiary carbocation, as it is the greater resonance contributor. However, that product is the less stable (kinetic) product. This product is tertiary and allylic. Therefore, it can reverse and react to give the more stable thermodynamic product, the most substituted alkene.

In the following example, protonation gives tertiary and secondary carbocations. The stability and reactivity differences are less than with tertiary and primary carbocations. Therefore more charge will be found on the secondary carbocation and the barrier to reaction at that carbon will be less. It also leads to the most stable alkene. Therefore, 1,2-addition leads to the kinetic and thermodynamic product.

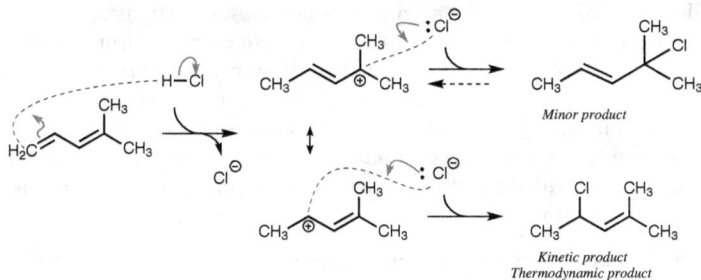

Minor product

Kinetic product
Thermodynamic product

11. Addition of bromine to 2-methyl-1,3-butadiene can give 3,4-dibromo-3-methylbut-1-ene, the kinetic product and 1,4-dibromo-2-methylbut-2-ene, the thermodynamic

product. If the reaction can be conducted at low temperatures, the kinetic products may be isolated.

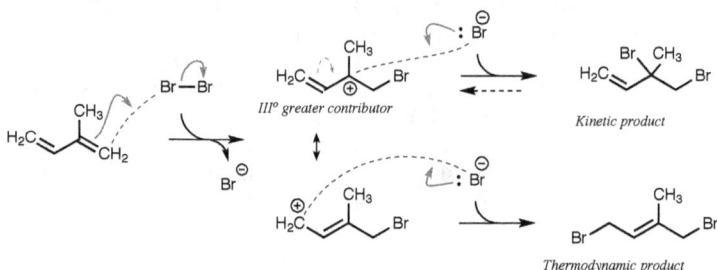

III° greater contributor

Kinetic product

Thermodynamic product

Bromination

12. Bromination of cyclohexene gives *trans*-1,2-dibromocyclohexane.

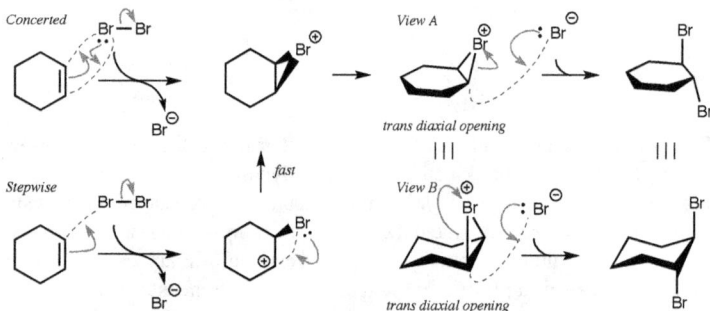

Concerted

Stepwise

fast

View A

trans diaxial opening

View B

trans diaxial opening

Many textbooks show bromination as a concerted reaction with three bonds forming and breaking. I prefer a stepwise mechanism. The reactivity is easily understood in a stepwise mechanism in which the alkene acts as the nucleophile and bromine the electrophile. This agrees with electron-rich alkenes reacting faster than electron poor ones because they take on a positive charge. An added step in the formation of a bromonium ion is only an electron movement and therefore is very fast.

In a concerted reaction, the addition may be considered a symmetry forbidden 2+2 addition. Also, although the alkene gains and looses electrons equally, the bromine must act as a nucleophile in that it donates more electrons than it accepts. I would predict little difference in alkene reactivity if no net charge is gained or lost by the alkene. Bromine is a strong electron withdrawing group and formation of a positively charged bromonium ion is counter to other reactions of bromine. I anticipate any electron movements can always occur much faster than atom movements and therefore a fast cyclization with the electrons of a neighboring bromine atom should not preclude a stepwise mechanism.

In the bromination of cyclohexene, it doesn't matter at which carbon the bromide enters to open the bromonium ion, as it is symmetrical. It only matters that the entering bromide is on the opposite side of the bond being broken in an S_N2-like reaction.

If an anchoring *t*-butyl group were present, then the diaxial product is formed upon opening the bromonium ion. The entering and leaving electrons can be aligned in an S_N2 like reaction as shown below. Note how the bromine is close to the entering trajectory for a diaxial product while they are quite distant for a diequatorial product.

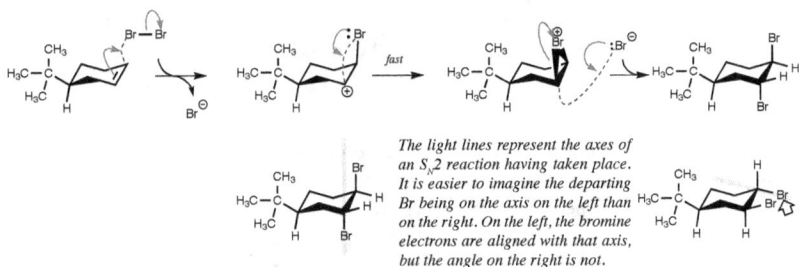

The light lines represent the axes of an S_N2 reaction having taken place. It is easier to imagine the departing Br being on the axis on the left than on the right. On the left, the bromine electrons are aligned with that axis, but the angle on the right is not.

13. Bromination of *trans*-2-butene gives (erythro) *(2R,3S)*- and *(2S,3R)*-2,3-dibromobutane. In this case, it is a *meso* product.

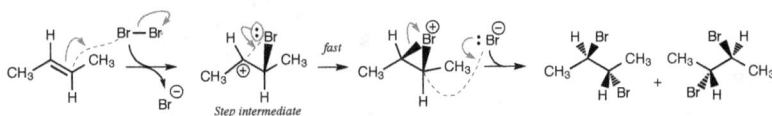

14. Bromination of *cis*-2-butene gives (threo) *(2R,3R)*- and *(2S,3S)*-2,3-dibromobutane.

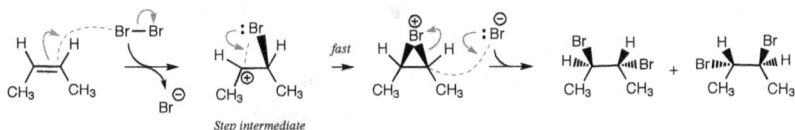

15. Bromination of methylcyclohexene gives *(1R,2R)*- and *(1S,2S)*-2-bromo-1-methylcyclohexanol. This reaction is stereo and regiospecific. In this case, the nucleophile (water) attacks the most substituted carbon and results in inversion. The stereochemical consequences of this reaction are the same as an S_N2 reaction. However, as a general rule, S_N2 reactions do not take place on tertiary carbons. How is this different? Generally, in an S_N2 reaction, the process of bond formation precedes bond cleavage. In this instance, you can think that unlike the S_N2 reaction, in which little bond cleavage has occurred to expose the nucleus to attack, the bond may be considered highly polarized and thus the carbon contains considerable positive charge. Therefore, the reaction is more S_N1 like than S_N2 like. The atom best able to support that charge is now the site of an S_N2-like reaction.

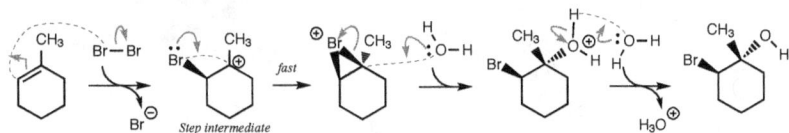

Alternately, you may think of it as though a three membered ring were not present (though **incorrect**). The reaction is then like an S_N1 reaction in which the neighboring atom and its non-bonded electrons blocks one face of the carbocation and results in attack from the opposite side.

The opening of the bromonium ion will again be *trans*-diaxial, see Example 5.12. The electrons attack the carbon nucleus from the opposite side of the leaving electrons of the bromonium ion. You may draw a Newman projection of a cyclohexane to see the stereochemistry of this reaction.

A bromination reaction in water results in some confusion. Even though bromide ion has a negative charge, it isn't the strongest base. Note in the equilibrium below, the equilibrium lies to the right. Therefore, water is a stronger base. Since water is the stronger base, it is plausible that its electrons can extend further and allow it to react faster with a bromonium ion than bromide ion.

Oxymercuration

16. Step 1, oxymercuration of 3-methyl-1-butene gives 3-methyl-2-butanol. This reaction is a good analog to the bromination reaction. In the reaction of an electrophilic mercury atom, the alkene will be the nucleophile. We might forget that mercury has ten valence electrons. Therefore, after it reacts with an alkene, mercury has electrons available to react with a neighboring positive charge, just as the bromine atom did.

In the carbocation intermediate, a hydrogen atom is staged for migration to give a tertiary carbocation, see Example 7. If the non-bonded electrons of the mercury did not react faster with the carbocation, then a rearrangement might take place. The mercury atom, with its electrons, prevents the rearrangement reaction from taking place.

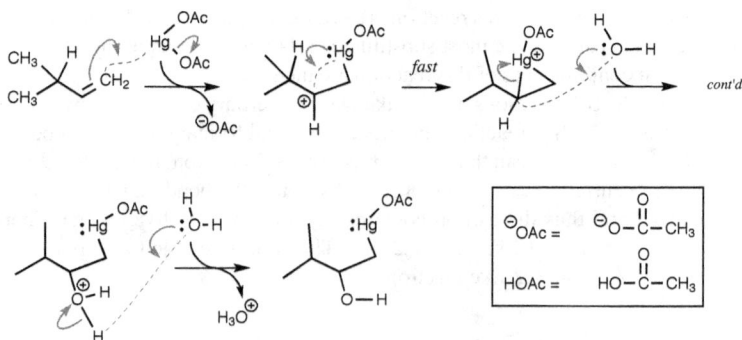

Step 2, reductive demercuration.

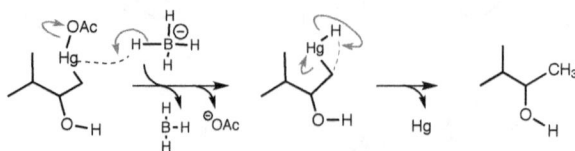

The mercury is removed by a borohydride reduction followed by loss of metallic mercury. The loss of mercury is reportedly a radical reaction. Two possibilities are written below. However, for expediency, I suggest you write the reaction as indicated above unless directed otherwise by your instructor.

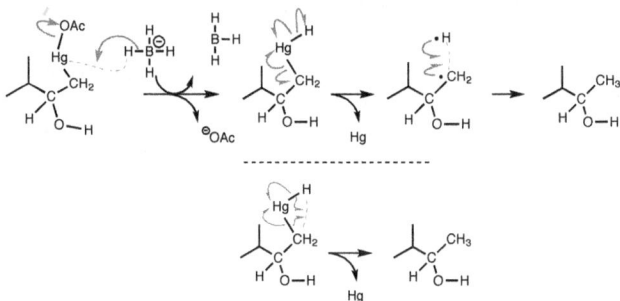

17. Step 1, oxymercuration of 1-methylcyclohexene gives 1-methoxy-1-methylcyclo-hexanane.

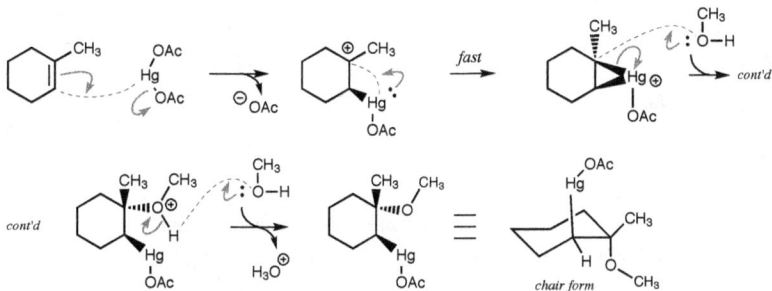

Step 2, reductive demercuration.

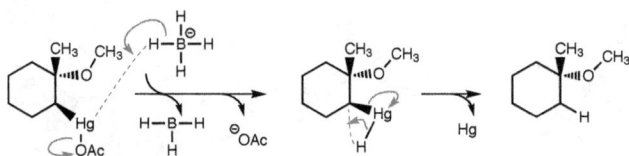

Hydroboration-Oxidation of Alkene

18. Hydroboration-oxidation of propene gives 1-propanol. Note how similar step one of this reaction is to the prior alkene reactions. Boron is an electrophile and reacts with the electrons from the alkene. The second stage of this reaction takes place before any atom movement can take place. The electrons from the negatively charged boron are donated to the carbocation. In this case, a proton is attached to the donated electrons.

I formerly indicated, to repeat the first reaction two times. While this is expedient, I found many students didn't understand what was happening in the repeat statement, thus the complete mechanism was added. In learning the mechanism, when it becomes clear to you that it is the same reaction being repeated three times, you can just write the first hydroboration and then indicate that the hydroboration step is repeated 2X.

Step 1, hydroboration. Each bracket represents one of three hydroboration steps.

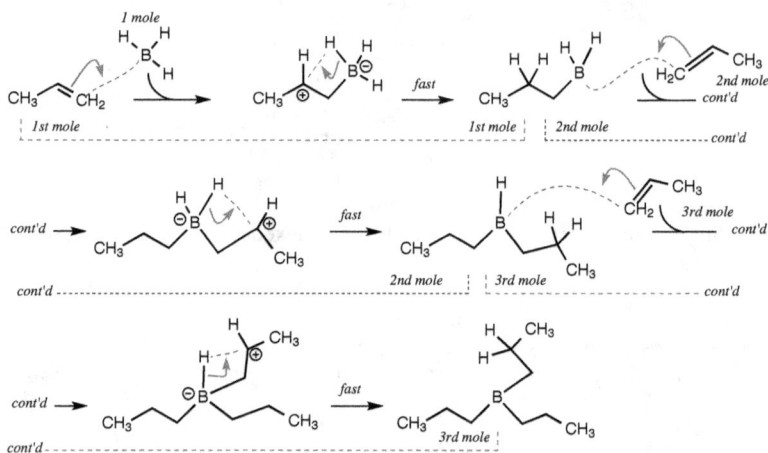

Step 2, oxidation. Each bracket represents one of three oxidation steps. Boron is still an electrophile except the nucleophile is now a hydroperoxy anion. Note, it is the same reaction repeated three times. You can just write the first oxidation step and then indicate that the oxidation step is repeated 2X.

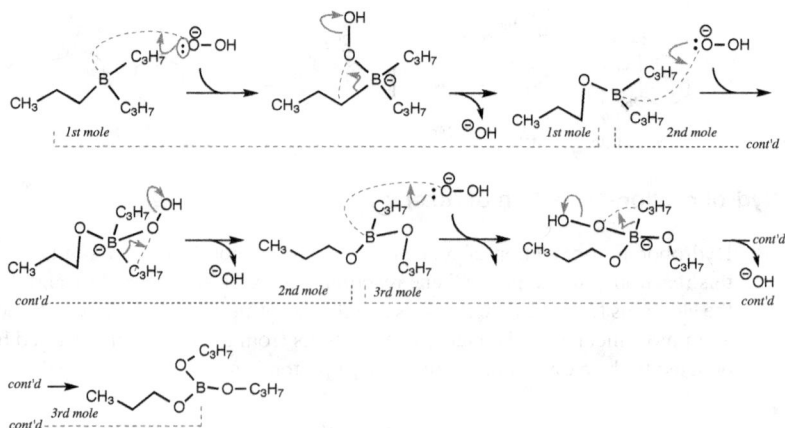

Step 3, borate ester hydrolysis to 1-propanol. Each bracket represents one of three hydrolysis steps.

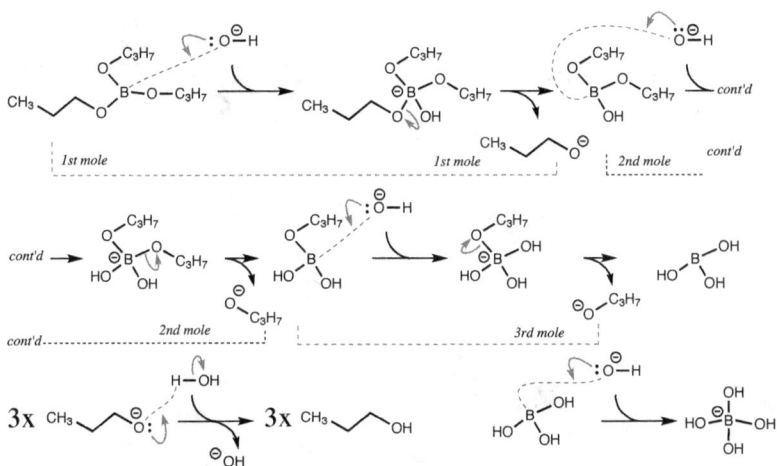

You may also use this alternate mechanism for the hydrolysis. I like the deprotonation to promote the loss of the alkoxide. However, the next step requires an attack on a negatively charged boron with an additional pair of electrons . The product of that addition is an sp³-hybridized boron with an alkoxide oxygen neighbor. While I did not use this mechanism, you may use it if your textbook or your instructor uses it.

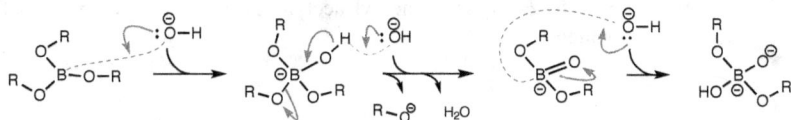

Stepwise versus Concerted Reactions

Because the addition of borane to an alkene is a *syn* addition, many books write it as a concerted reaction. I have not done so for two reasons. One, it appears to be a symmetry forbidden 2 + 2 reaction. The symmetry rules can be used to explain why alkenes and HBr do not thermally add to an alkene in a concerted reaction, see Example 1. The same principle should apply here. Two, if the reaction were concerted, then charges should be minimized and therefore the selectivity of the reaction should be reduced.

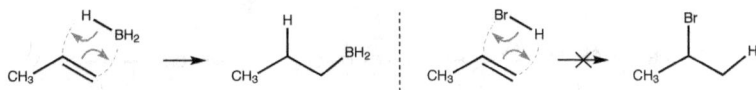

I suggest a definition of a concerted reaction as any electron movements that occur faster than bond rotation are concerted. Therefore a reaction may be concerted and have additional steps provided they occur faster than atom movements. In this case, the stepwise mechanism preserves the reactivity of an alkene as a nucleophile and borane as an electrophile. This is consistent with the majority of boron and alkene reactions. The movement of electrons to form a bond can be expected to occur faster than atom movements and thus be consistent with a concerted reaction.

19. Hydroboration/oxidation of 1-methylcyclohexene gives *trans*-2-methylcyclo-hexanol.

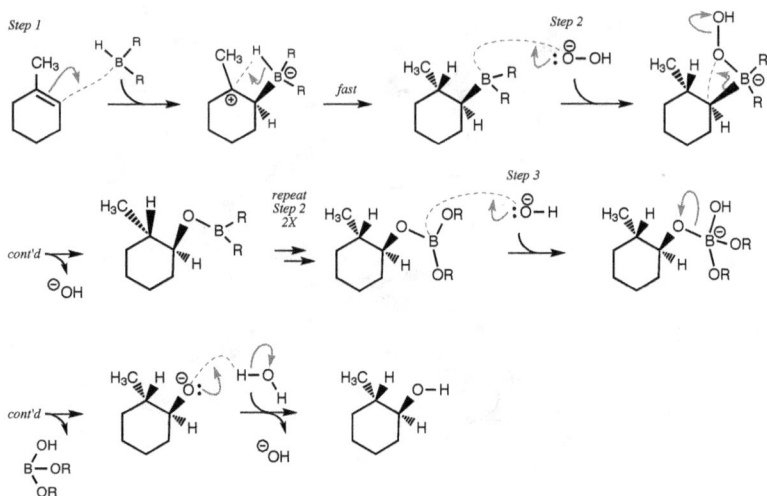

Carbon-Carbon Triple Bond Electrophilic Reactions

Addition to an Internal Acetylene

20. Addition of HCl to 2-butyne (dimethyl acetylene) gives (*E*)-2-chloro-2-butene and 2,2-dichlorobutane.

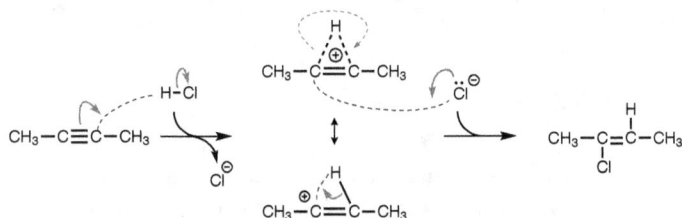

2nd Equivalent of HCl

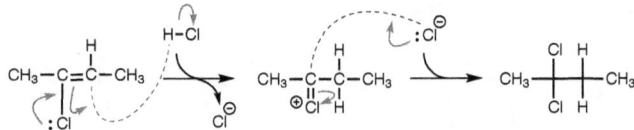

The three membered ring immediate above contains a kind of misrepresentation. The addition of HCl to the acetylene is an example of a three-centered two-electron bond. As written, it implies two bonds to hydrogen. I prefer the representation shown below. If the proton added to a pair of electrons, an intermediate such as on the left would result. The charge of the nuclei would be unchanged, but an imbalance would exist. Because the acetylene carbons are sp-hybridized, one less bond or carbon is available to donate electrons and stabilize a carbocation. As a result, you may think the electrons remain attached to the acetylene carbons. When a nucleophile approaches one of the carbons, it will only be upon addition that bond cleavage occurs. I prefer this model to explain why addition of HCl gives mainly the *trans*-product.

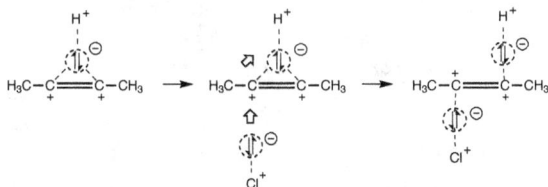

If we compare the acetylene reaction with addition of HCl to propene, the additional carbon bond to the carbocation gives it greater stability. It results in the most stable carbocation and is consistent with a Markovnikov addition. It can also be used to explain why addition of HCl and HBr to alkenes can gives high levels of *anti*-addition.

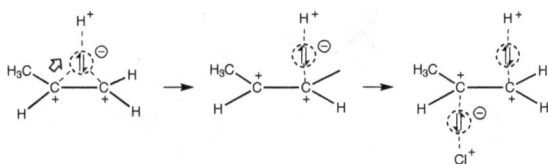

21. Addition of bromine to ethynylcyclopentane gives *(E)*-(1,2-dibromovinyl)cyclopentane. The three-membered bromonium ion does contain three bonds and should not be confused with the proposed model for HCl addition.

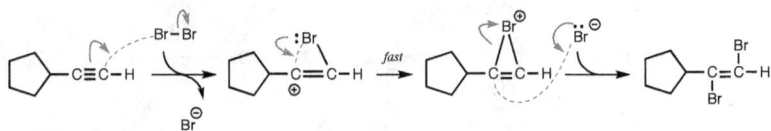

Addition of a second equivalent of bromine gives (1,1,2,2-tetrabromoethyl)cyclopentane.. The bromine atoms reduce the reactivity of the alkene. Therefore it is possible to perform a stepwise addition of one or two equivalents of bromine.

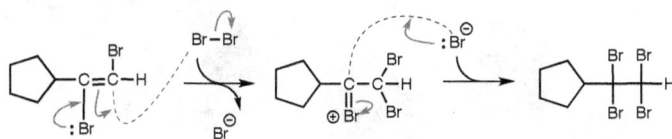

22. Sulfuric acid catalyzed hydration of 1-propynylbenzene gives 1-phenyl-1-propanone. This reaction starts similarly to Example 20. In this case, it doesn't matter if it opens *trans* as the stereochemistry is eventually lost. Sulfuric acid is a good acid to use because its conjugate base, bisulfate ion, is a weak nucleophile, its negative charge is distributed over several atoms, and does not compete for product formation. This reaction requires an additional carbon protonation step to give a ketone as the final product. *In the last deprotonation step, which proton is more acidic, the one on the oxygen (to give the ketone) or the one on the carbon (to regenerate the enol)?*

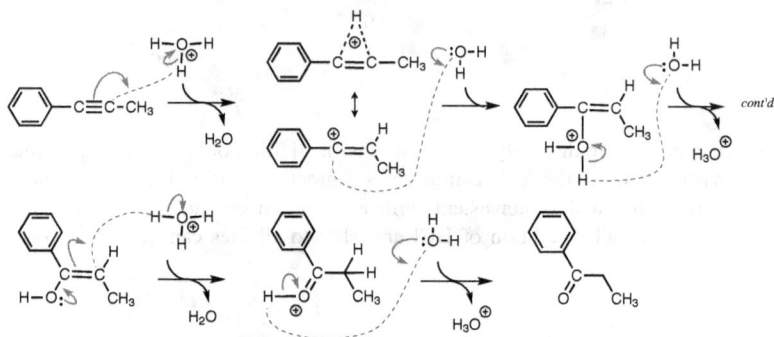

Addition to a Terminal Acetylene

23. Mercury catalyzed hydration of propyne (methyl acetylene) gives 2-propanone (acetone). With an internal acetylene, two carbons are attached to the acetylene, they donate electrons to the acetylene, and protonation can occur directly. With a terminal acetylene, mercury is needed to convert it into a ketone. The oxymercuration of a terminal alkyne parallels the alkene reaction. However, the product would be an alkenylalcohol. This alkenylalcohol can tautomerize. During the tautomerization, mercury gives up its electrons to reform mercuric ion and a ketone.

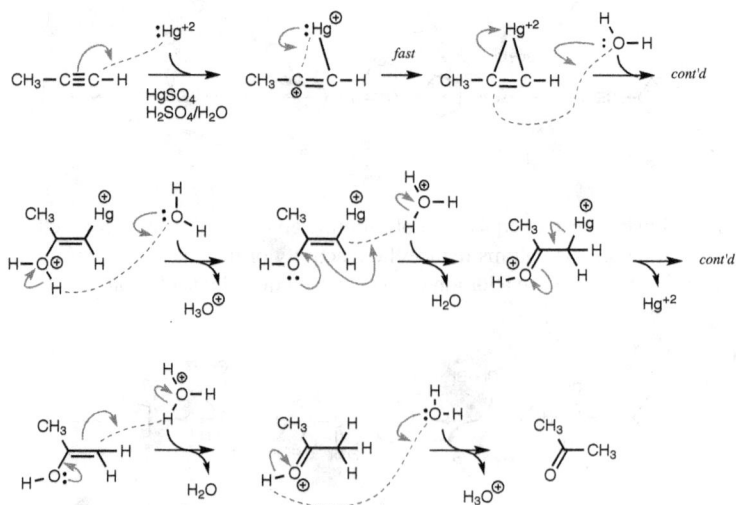

Disiamylborane Hydroboration–Oxidation of an Acetylene

24. Hydroboration-oxidation of phenylacetylene with disiamylborane gives phenylacet-aldehyde. A hindered borane is used to avoid over reduction of the acetylene. The intermediate product is an alkenylalcohol (enol) which tautomerizes to an aldehyde.

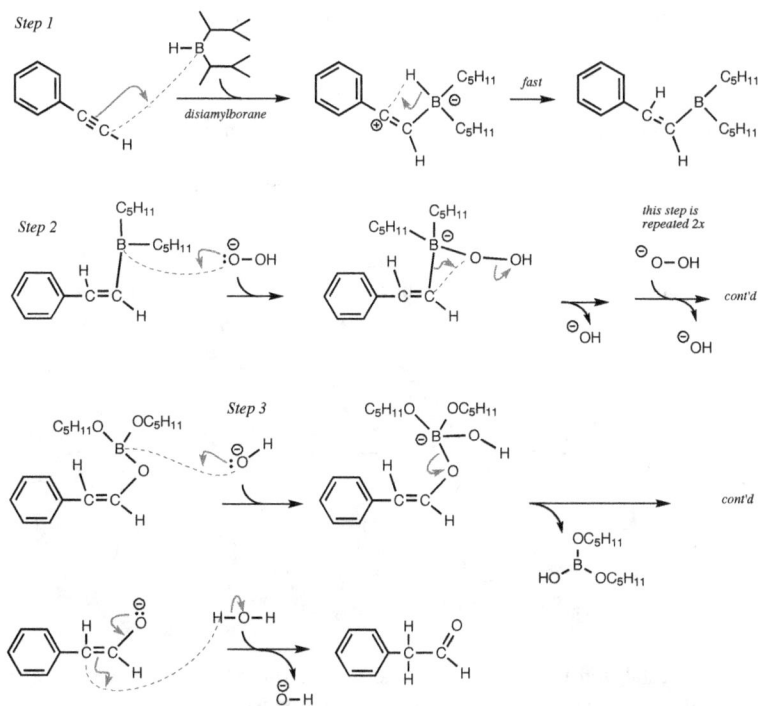

6 — Rearrangement Reactions

Baeyer-Villiger Oxidation

1. Acid catalyzed Baeyer-Villiger oxidation of 2,2-dimethylcyclopentanone with peracetic acid. A key to predicting Baeyer-Villiger oxidation products is being able to predict which group will migrate. The order of migration is hydrogen > tertiary alkyl > secondary alkyl > phenyl > primary alkyl > methyl, but there are many exceptions based on the peracid used, reaction conditions, and stereochemistry. The electronic effect or stereochemistry is similar to elimination reactions with the migrating electrons preferring an *anti* arrangement. The migrating group can be described as the one best able donate electrons. Except for hydrogen (Examples 3 and 7), which becomes lost, a large carbon-electron pair distance increases migratory ability (tertiary>secondary>primary, see Chapter 1).

    I have written Baeyer-Villiger rearrangement reactions for two types of conditions, acid and base catalyzed.* Peracetic acid is a commonly used peracid and typically it is stabilized with sulfuric acid. Therefore it is an example of an acid catalyzed reaction.

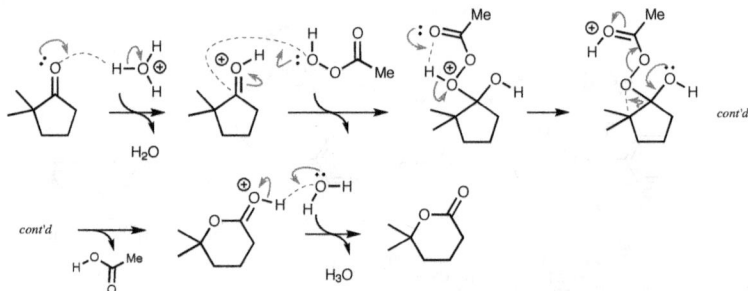

2. Acid catalyzed Baeyer-Villiger oxidation of *o*-methoxyacetophenone with peracetic acid. This example makes use of the non-bonded electrons of the ether to facilitate the rearrangement.

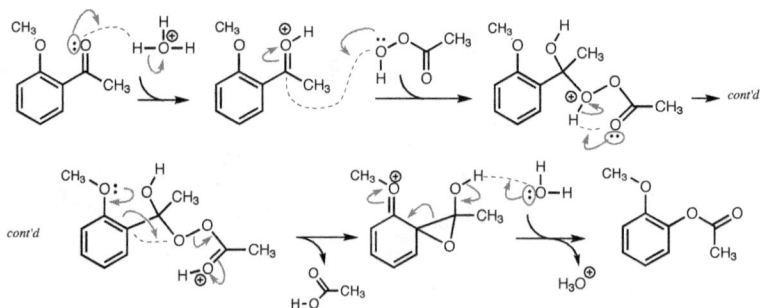

    I had written two additional phenyl rearrangement mechanisms. The phenyl group migrates in either rearrangement. The first version better shows why an electron-donating group would migrate because it requires a positive charge be placed upon the ring. In the second version, while we might expect the non-bonded electrons

---

* A Baeyer-Villiger oxidation requires an addition to a carbonyl group. I have described different scenarios for an addition in Chapter 8. Acidic and basic additions are two of the options.

of oxygen to participate (like a resonance structure), the electron pushing from the oxygen does not make any demands of the phenyl ring.

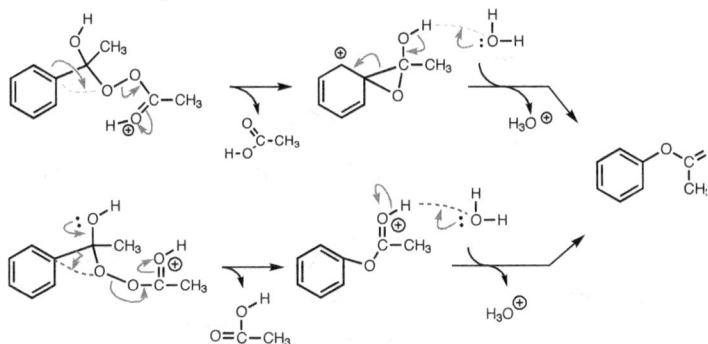

3. Acid catalyzed Baeyer-Villiger oxidation of benzaldehyde with peracetic acid gives benzoic acid. This reaction may also be thought of as an oxidation.

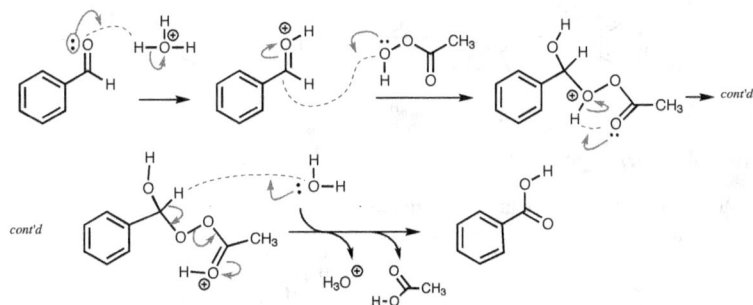

4. Baeyer-Villiger oxidation of a benzophenone with trifluoroperacetic acid. If acidic conditions should be avoided, you may find a buffer has been used (peracid plus a weak base like bicarbonate). Many of those examples use trifluoroperacetic acid. This example has two aromatic rings. The more electron rich ring migrates most quickly.

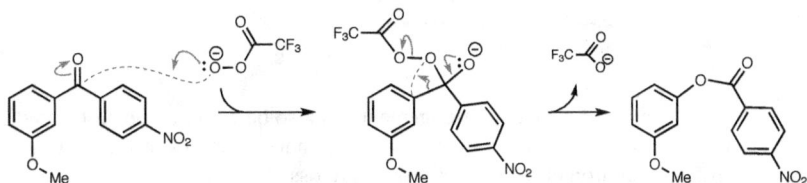

5. Baeyer-Villiger oxidation of o-methoxyacetophenone with m-chloroperoxybenzoic acid (MCPBA).

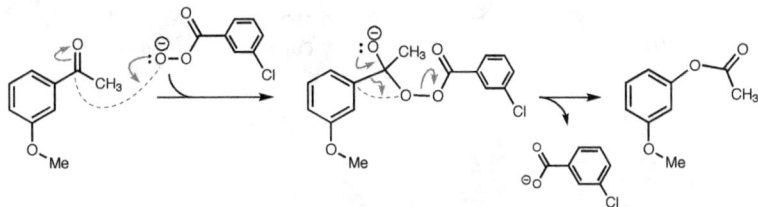

6. Baeyer-Villiger oxidation of bicyclic phenyl ketone with trifluoroperacetic acid.

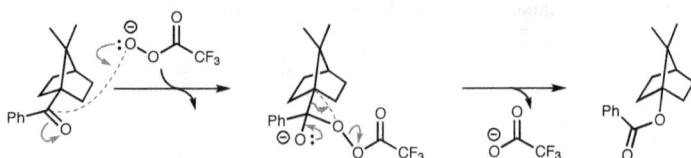

7. Baeyer-Villiger oxidation of *p*-chlorobenzaldehyde with peracetic acid gives *p*-chlorobenzoic acid.

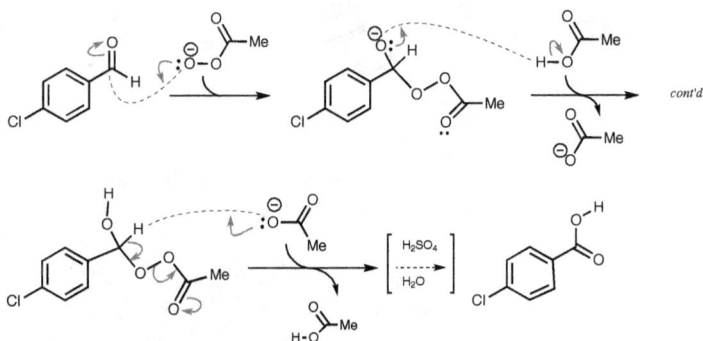

cont'd

Pinacol Rearrangement

8. Rearrangement of pinacol to pinacolone (methyl *t*-butyl ketone or 3,3-dimethyl-2-butanone).

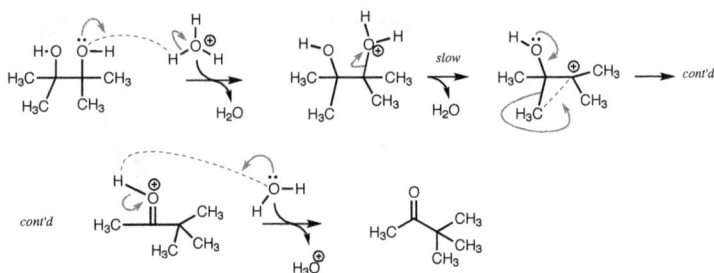

cont'd

A more concerted pinacol rearrangement can also be written. The protonated alcohol can be rearranged to the product. The non-bonded electrons of the adjacent oxygen facilitate migration of the methyl group and loss of water.

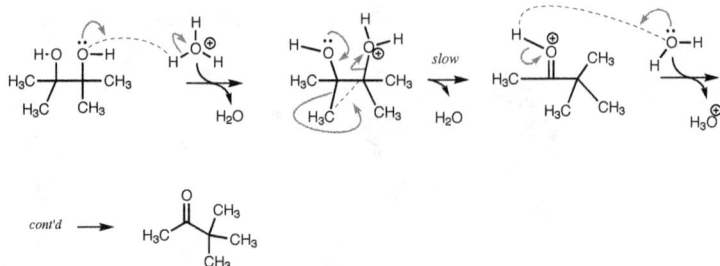

cont'd

9. Rearrangement of 1,2-dimethyl-1,2-cyclohexanediol with acid to a methyl ketone.

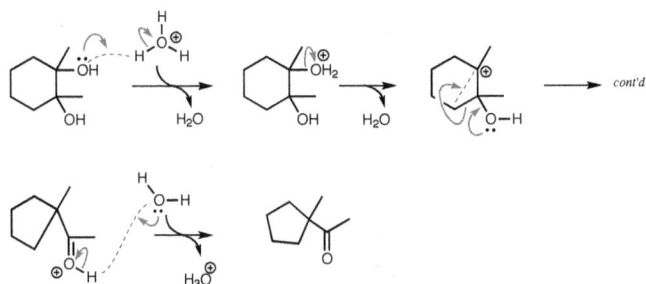

Benzilic Acid Rearrangement

10. Reaction of benzil with hydroxide gives benzilic acid after rearrangement. Now, if you are beginning to look at reactions as which atoms are donating or pushing and which are pulling, then this rearrangement is like the Baeyer-Villiger oxidations, but rather than breaking an O-O bond, this breaks a C-O double bond (second structure).

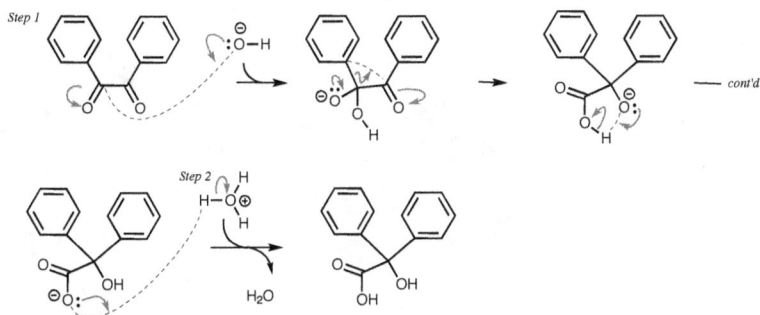

Dakin Reaction

11. Reaction of an o- or p-hydroxybenzaldehyde with basic hydrogen peroxide gives a phenol. This reaction is narrow in scope, but does not result in an oxidation of the aldehyde. The p-hydroxy group probably reduces the ability of the intermediate to accept electrons from a proton loss. Therefore, rather than lose the proton, a rearrangement takes place. Optionally, the reaction might have involved the phenoxide anion which would further increase the C-H carbon-electron distance and further reduce the acidity of the hydrogen.

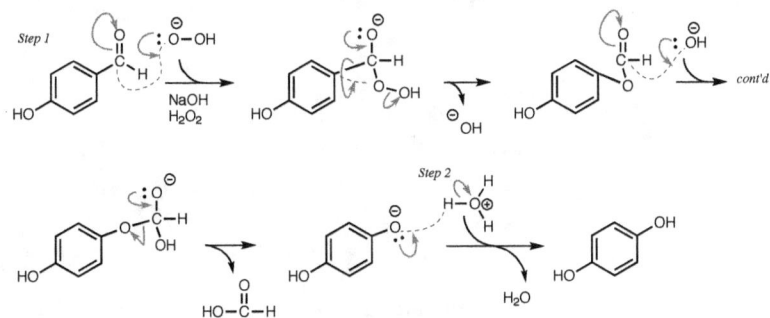

A Handbook of Organic Chemistry Mechanisms • 57

Acetone from Cumene

12. A reaction of isopropylbenzene (cumene) to give acetone and phenol. This is the commercial production of acetone and phenol. It is carried out on a very large scale. Air or oxygen is used in the first step. Then the reaction simply requires a catalytic amount of an acid.

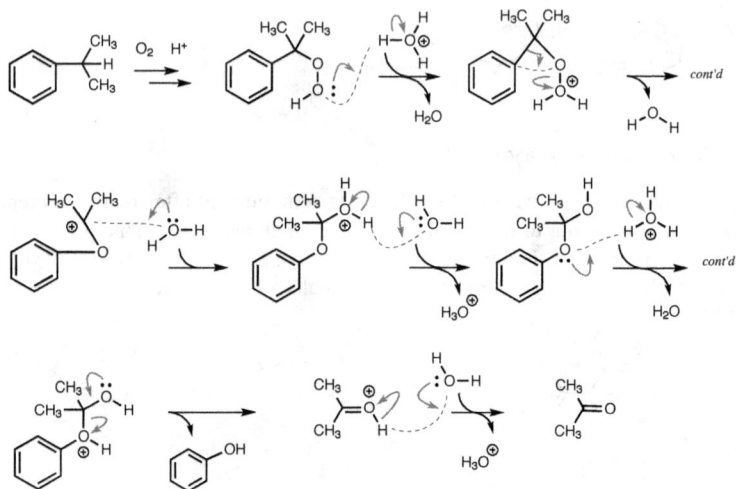

7 — Electrocyclic Reactions

Diels Alder Reactions

The Diels-Alder reaction is simple, but errors in connecting the bonds are common. Many books use dashed lines to indicate where new bonds should form. In addition, I suggest you label the atoms of the reactants and products, and check that the bonds correspond with the labels. For bicyclic products, I suggest breaking this into two operations. First, complete the bonds for the product. Then, convert the two-dimensional structure into a correct three-dimensional structure. I can make an error if trying to do both at the same time.

1. A Diels-Alder reaction between 1,3-butadiene and 2-propenal (acrolein).

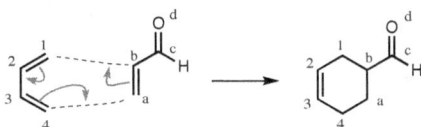

2. A Diels-Alder reaction between 1,3-cyclopentadiene and (E)-2-butenal (*trans*-croton-aldehyde). The stereochemistry of the double bond is preserved in the formation of the product. This is one of the great features of the Diels-Alder reaction. In this example, four stereocenters have been set in one reaction.

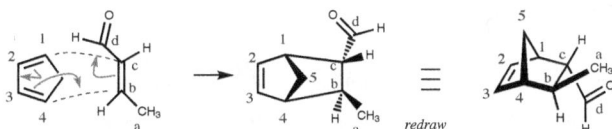

This is the same example as above. It is more difficult to negotiate the atom movements, congestion, and drawing the final product all at once. I wrote the upper example in two steps as it is easier to visualize and convert it to the bicyclic product. This example may portray the the electron-withdrawing group of the dienophile overlapping with the diene. This represents the preferred orientation of the dienophile also.

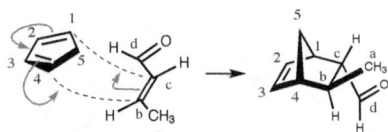

3. The Diels-Alder dimer of cyclopentadiene.

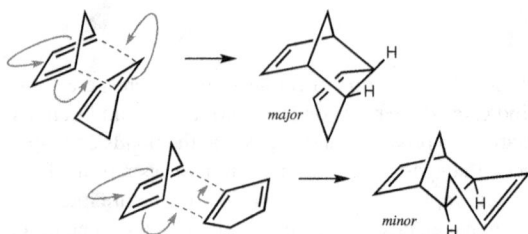

major

minor

4. A Diels-Alder reaction between 1,3-butadiene and methyl (Z)-2-butenoate. Note the dienophile stereochemistry is preserved.

5. A Diels-Alder reaction between 2-methyl-1,3-butadiene and N-methylmaleimide.

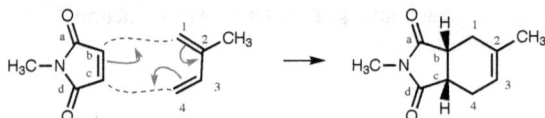

6. A Diels-Alder reaction between furan and but-3-en-2-one (methyl vinyl ketone, MVK).

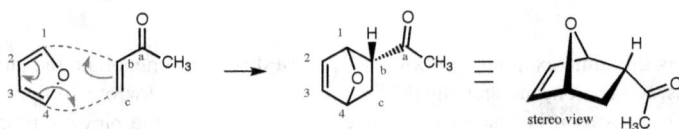

stereo view

7. A Diels-Alder reaction between cyclopentadiene and dimethyl acetylenedicarboxylate.

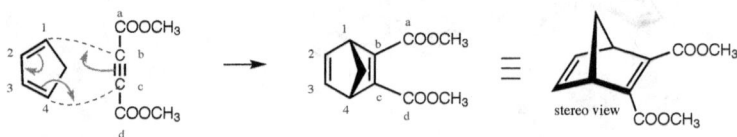

stereo view

8. A reverse-forward Diels-Alder reaction between cyclopentadiene and maleic anhy-dride. Yes, these reactions can reverse. That is the source of cyclopentadiene. It is then free to react with a more reactive dienophile than cyclopentadiene itself.

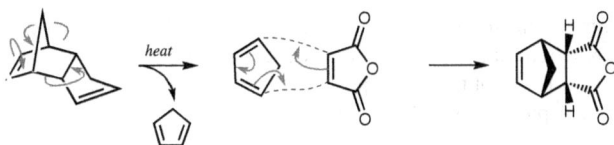

9. A reverse-forward Diels-Alder reaction between butadiene (sulfone) and maleic anhy-dride.

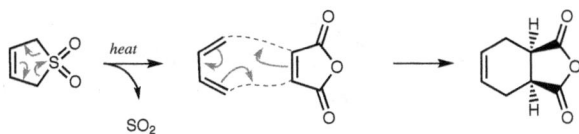

10. A Diels-Alder reaction between (2E,4E)-hexa-2,4-diene and maleic anhydride. Again, notice that four stereocenters are created in this reaction.

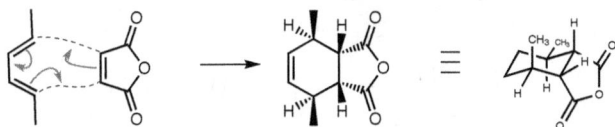

11. A Diels-Alder reaction of 1,3-cyclohexadiene and 2-butenal (crotonaldehyde).

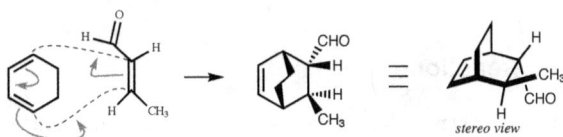

12. A Diels-Alder reaction between 1-methoxy-1,3-butadiene and methyl vinyl ketone gives a major product. The Diels Alder reaction is generally regarded as a concerted electrocyclic reaction. This example blurs the lines for concerted reactions. Writing it as an ionic reaction would predict the correct regiochemistry. However, it is NOT a stepwise ionic reaction.

The boxed structure is the preferred product. It matches a Coulombic attraction of the charges in the resonance structures (which might contribute to the transition state).

The ionic contributions are drawn in the dashed boxes. The resonance structure isolate the bond formation into two independent steps. Those resonance structures would hinder formation of the other isomer. In a completely concerted reaction, there should be little build up of charge and both products could result.

The true nature of this reaction contains characteristics of being an ionic reaction and a concerted reaction. Some electron movement characterized by the resonance structures must occur to orient the major product yet retain enough concerted character to lead to the minor product as well.

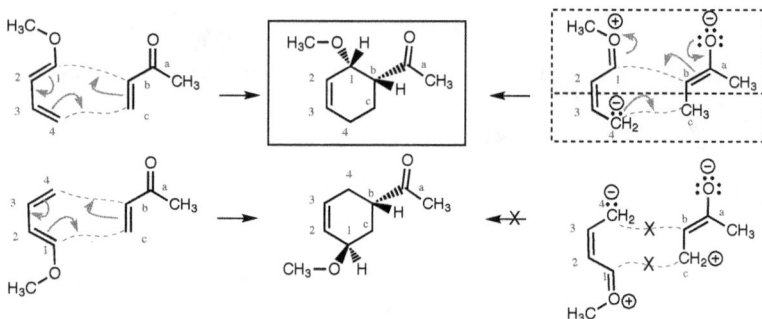

The resonance structures of the diene and dienophile show the charges are complementary for the formation of the major product.

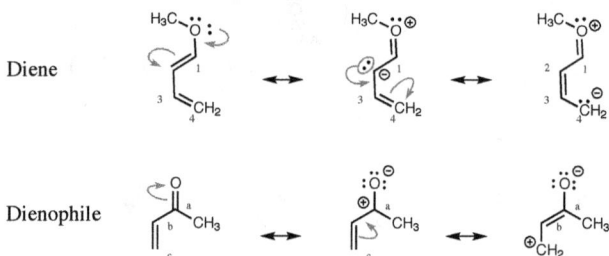

Other Electrocyclic Reactions

13. A 3+2* cycloaddition between cyclopentene and benzonitrile oxide. It fits the general model for electrocyclic reactions and extend the model. Three pairs of electrons (curved arrows) move to complete the formation of product. It is the number of electrons that move in a Diels-Alder reaction.

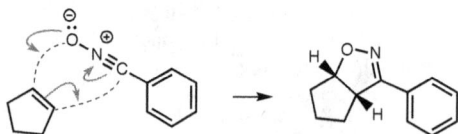

---

* The numbers refer to the numbers of atoms in each ring forming element. The other Diels-Alder reactions are also referred to as 4+2 electrocyclic addition reactions.

14. A Claisen rearrangement (electrocyclic) reaction to transfer a group from oxygen to carbon. It is an exaggeration that hydronium ion is necessary for the tautomerization to take place. The product phenol is a more likely acid that catalyzes the tautomerization. Drawing the phenol as the acid was awkward.

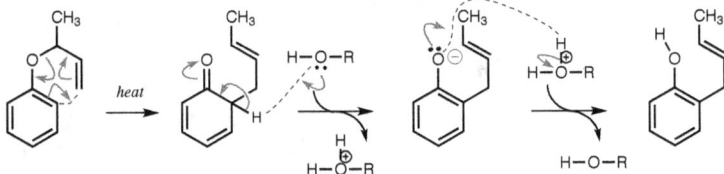

15. A double Claisen rearrangement (electrocyclic) reaction to transfer a group from oxygen to carbon to another carbon. It is the same number of electrons that move in each electrocyclic step.

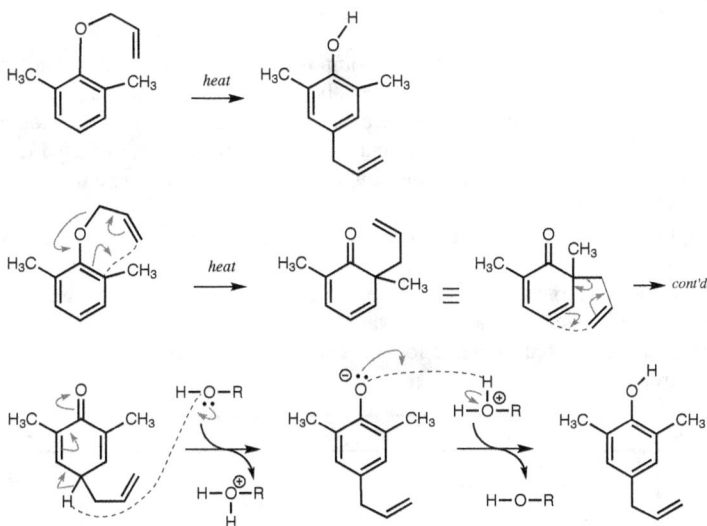

8 — Carbonyl Addition and Addition-Elimination Reactions

Additions by C, N, and O. Addition by hydrogen nucleophiles are discussed in Chapter 11, Reduction.

Addition to Carbonyl Groups

For us to understand carbonyl addition reactions, we need to compare carbonyl addition reactions to simple acid-base reactions. Upon addition of HCl to water, an equilibrium is established in which the concentrations and reactivity of the species are used to determine the equilibrium. Thus, the equilibrium between HCl and water shifts to the right to form the weaker acid, hydronium ion, see Chapter 2.

$$HCl \quad + \quad H_2O \quad \rightleftarrows \quad \overset{\ominus}{Cl} \quad + \quad \overset{\oplus}{H_3O}$$
$$\text{-7} \hspace{6.5cm} \text{-1.7}$$

For an addition to a carbonyl group, the composition of the equilibrium can also be analyzed by acid-base properties. In Example A below, if a nucleophile adds to a protonated carbonyl group, the product is a weaker acid, therefore the equilibrium* should favor an addition. In Example B, an anion adds and the equilibrium can be estimated from the pK_a of the conjugate acids. If $X = CH_2$, this could represent a Grignard reaction and product is an alkoxide. The difference in pK_a of ethane (50) and an alcohol (*ca* 18) favors the alcohol. In Example C, a neutral atom has added to form a zwitterion intermediate. If $X = N$, the electrons are more available and the difference in basicity of the starting amine and the zwitterion is not as large as if $X = O$. Also, the geminal cation will decrease the pK_a of the alcohol (the conjugate acid). Finally, Example D is a completely concerted reaction. No net charges are generated to create local acids or bases. However, this mechanism has a greater entropic requirement.

Example	Reaction		
A			
B			
C			

---

* I am only considering the displayed equilibrium. A simple proton transfer to the X-atom can also occur and decrease the concentration of protonated carbonyl compound.

| D | 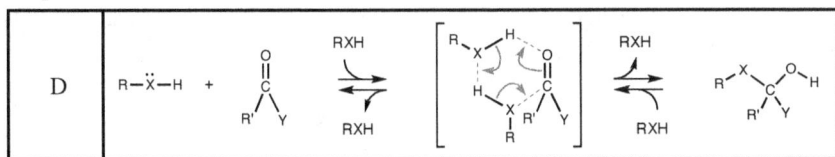 |

You may also encounter a variant of Example C in which the intermediate I have drawn is a transition state which transfers a proton via a four-membered ring. While this mechanism is present in some textbooks, I am troubled by a lack of precedent for this proton transfer. In a normal hydrogen bond, the preferred bond angle is 180°. Variations from 180° are commonly found in six-membered rings, see Chapter 9.8 and five-membered rings, see Chapter 4.18 and 19, and Chapter 12.7 and 8.

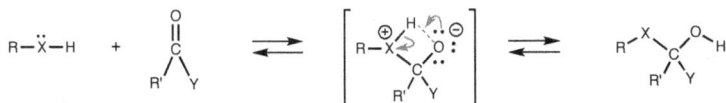

While the four-membered ring is expedient and avoids a zwitterionic intermediate, I am skeptical sufficient experimental data exists to support it. In the normal hydrogen bond, the electron-electron repulsion forces the nuclei to be linear. While smaller angles are present in six and five-membered rings, a continued decrease in bond angle increases the electron-electron repulsion exponentially as predicted by Coulomb's Law. This could be compensated for with a a large nucleus, see Chapter 8.5.[†] A larger nucleus can attract electrons and mitigate their repulsion. However, I have resisted writing any examples of proton transfers via four-membered ring intermediates.

Since most proton transfer reactions take place in protic solvents, intermolecular proton transfer are readily available. An alternate to the concerted and intramolecular proton transfer is an intermolecular proton transfer as shown below. This variation should have a lower entropic requirement and simply extends the life of the zwitterion intermediate.

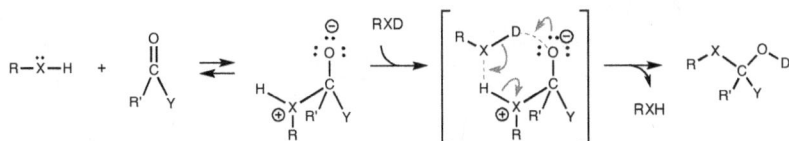

The final acid-base considerations are in the conversion of a geminal zwitterion of Example C to a neutral intermediate. If the reaction conditions are at a low pH, then pro-

---

[†] Baldwin addressed the electronic demands in ring formation reactions in Baldwin's Rules. With larger atoms, the leaving electrons can maintain an S_N2 like relationship. A better electronic analogy is S_Ni substitution reactions. These contain the same types of repulsive forces. Those examples are unclear whether the lack of inversion occurs is due to same side attack or caged ions.

tonation should precede deprotonation. If the reactions conditions are at a high pH, then deprotonaton should precede protonation.

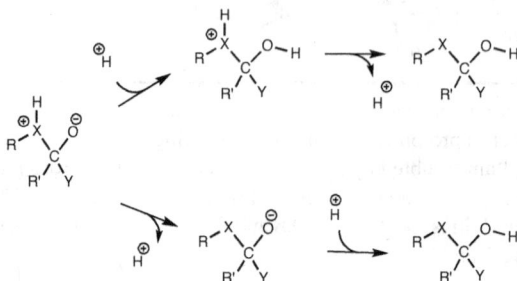

Through this chapter (and others), we must consider which intermediates might be present in a reaction with a carbonyl group. We must consider how the reaction conditions determine which intermediates might be present. If a reaction is acid catalyzed, then the mechanism of Example A may predominate. If the reaction is basic, then Example B may predominate. However, many reactions will be more difficult to determine which mechanism is operating. We must consider Example C in which two neutral compounds react together to give ionic intermediates. Then which proton transfer steps might be occurring to complete the reaction? Finally, Example D should also be considered. This example is completely concerted with low charge gradients but higher entropic demands.

I cannot predict which intermediate (or analog) might be present in any reaction. What you should note is that these mechanisms are schemes for proton transfers. You should ask yourself which scheme is the most compatible with the reaction conditions. If a different intermediate is used, it may or may not affect your ability to predict the product of a reaction. A mechanism is an attempt to rationalize how a reaction might take place. The objective of that rationalization is to make the chemistry logical.

I don't want anyone to think any mechanism that arrives at the product is acceptable. I have tried to write correct mechanisms as they teach us how reactions do take place. If the mechanisms are correct, they also teach us how other reaction should take place because those examples reveal the properties of the electrons in their various environments.

Grignard Addition to a Carbonyl Group

1. Addition of methyl magnesium bromide to cyclohexanone gives 1-methylcyclohexanol.

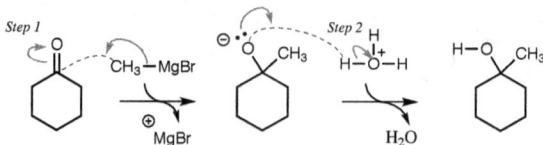

Grignard Formation

Grignard reagents are formed from magnesium metal and an alkyl or aryl halide. The ease of their formation is I>Br>Cl. Magnesium is thought to donate electrons in a single electron transfer process. The product of this reaction is methyl magnesium iodide or methyl

Grignard. Again, this reagent can react as though it were an anion. Grignard reagents are very reactive and must be protected from water. Anhydrous ether is a very common solvent used with Grignard reagents.

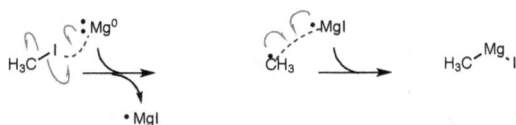

2. Addition of a Grignard reagent to acetaldehyde gives 6-methyl-2-heptanol. You may find organic compounds are written left to right as shown. When they react, the atom to atom connectivity must be maintained. This is a good example where numbering, lettering or numbering and lettering the reactants and products is useful.

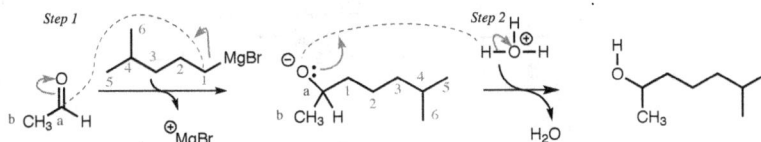

Alkyllithium Addition to a Carbonyl Group

3. Addition of ethyllithium to benzaldehyde gives 1-phenylpropanol.

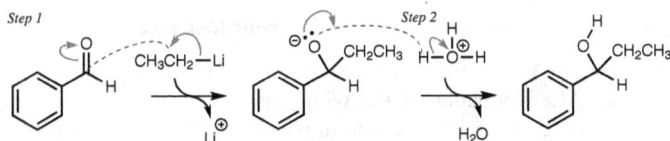

Alkyllithium Formation

Organolithium reagents are similar to Grignard reagents in their formation and reactions. Lithium has an electron that it donates to form a radical. Radicals are very reactive and thus reacts with a second equivalent of lithium metal to form the organolithium reagent.

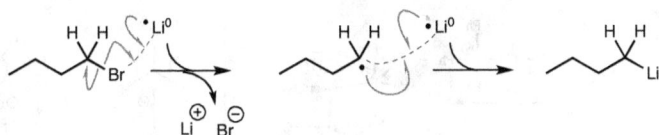

4. Addition of propynyllithium to acetone gives 2-methylpent-3-yn-2-ol. See Chapter 2.14 for the formation of an acetylide anion.

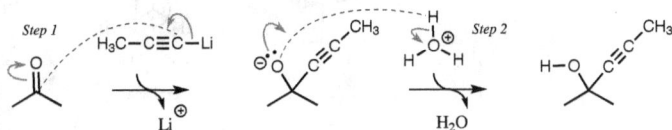

Wittig Reaction

5. Wittig reaction, Step 1, formation of Wittig reagent.

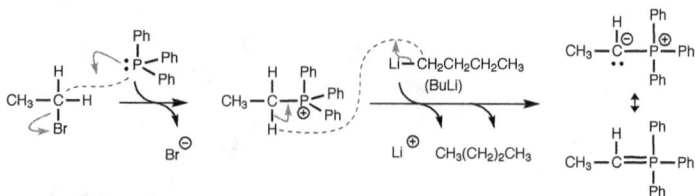

Step 2, reaction with benzaldehyde

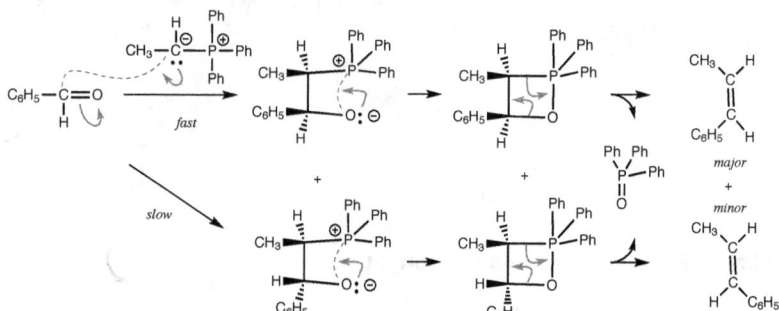

fast

slow

major

minor

Wittig Reaction and Horner-Wadsworth-Emmons Reaction

The Wittig and Horner-Wadsworth-Emmons variation of the Wittig reaction are comple-mentary reactions. After formation of the Wittig reagent (Step 1), the addition (of the ylide) to an aldehyde gives an intermediate in which the more stable conformation pre-dominates. This has been drawn conventionally and as the corresponding Newman pro-jection (see box). This conformation must undergo a bond rotation in order to bring the phosphorus and oxygen atoms together. That rotation gives the conformation that results in a *cis*-alkene. Therefore, the Wittig reaction stereochemistry is determined by the fast addition to the carbonyl group.

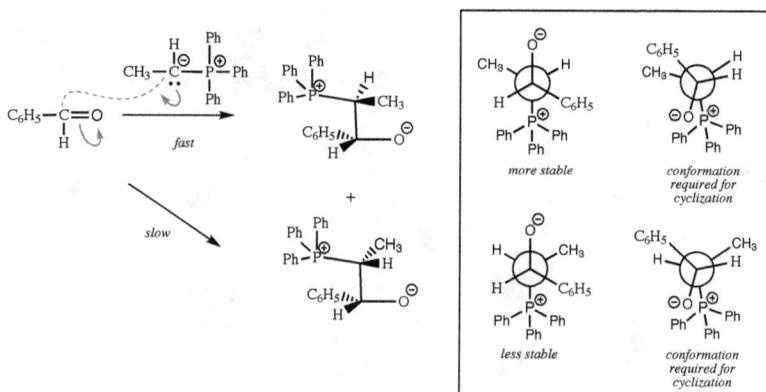

fast

slow

more stable

conformation required for cyclization

less stable

conformation required for cyclization

6.   Wittig reaction, Horner-Wadsworth-Emmons modification, Step 1, Arbusov reaction.

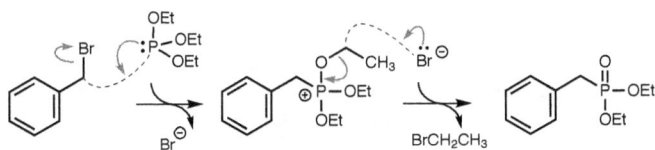

Step 2, reaction with benzaldehyde

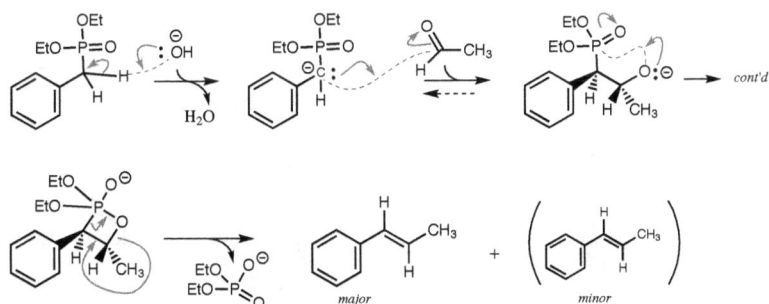

For the Horner-Wadsworth-Emmons variation of the Wittig reaction, a very similar set of structures can be written. The difference is the initial H-W-E-Wittig addition reaction is reversible. Because the reaction can reverse, the formation of a *cis*-alkene from the higher energy eclipsed conformation (the marked structure) is unfavorable. When the reaction reverses, the slower forming addition product can form and it can adopt the lower energy eclipsed conformation necessary for the cyclization-elimination step. This results in a *trans*-alkene.

This is a general phenomena. If stabilized Wittig reagents are used, they too can reverse to form thermodynamic products. I have not discussed other factors such as solvent and coordinating ions that can also become factors in determining the products.

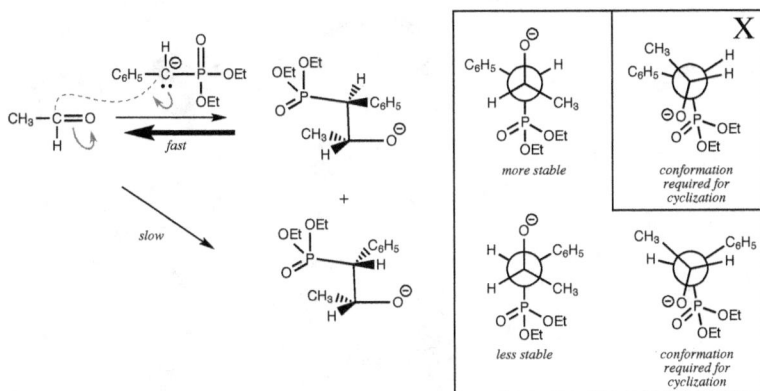

Addition-Elimination Reactions (Reversible Additions)

Ketal Formation and Hydrolysis

7. An acid catalyzed reaction of cyclohexanone with methanol gives the ketal. The formation of a ketal and its hydrolysis have been written with single forward reaction arrows. However, these reactions are reversible. The formation of products are controlled by their equilibria (Le Chatelier's Principle). If water is added, the equilibrium favors hydrolysis, and if water is removed, the equilibrium favors ketal formation.

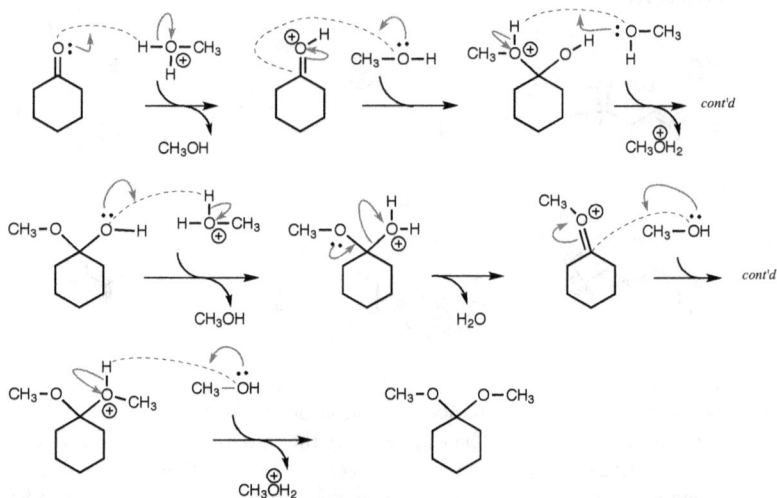

8. An acid catalyzed hydrolysis of a dioxolane acetal gives benzaldehyde. As stated above, the reaction is reversible. The difference in whether the product is an aldehyde or ketone is whether water is being added (for an hydrolysis) or removed (for ketal or acetal formation).

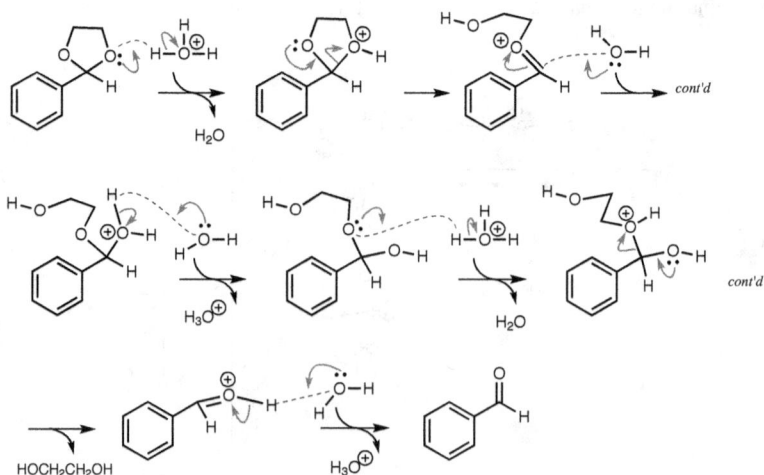

Oxime Formation

9. Formation of the oxime of cyclohexanone. This is a common mechanism for other primary amines also. The OH group can be replaced with an alkyl group, NH_2, or NRR' to produce a variety of C=N products. The mechanistic steps are the same or very similar. The acidity of the nitrogen atoms can affect the order and acids present in proton transfer reactions. For an oxime, the optimal pH is approximately 4, while the imine formation of a Wolff-Kishner reaction (Chapter 11.12) occurs under strongly basic conditions.

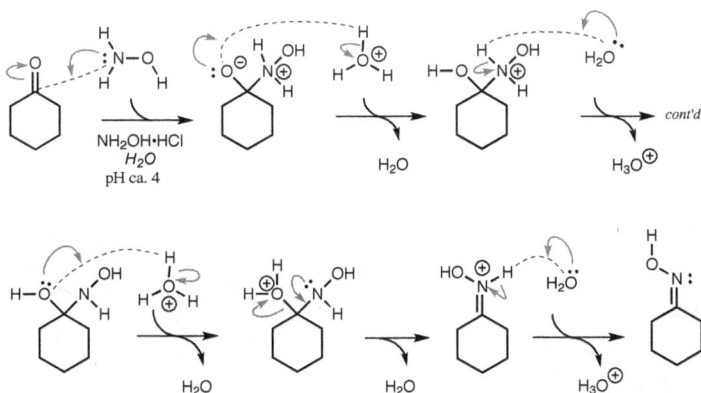

Other Additions to a Carbonyl Group

10. Formation of the cyanohydrin (2-hydroxy-2-methylpropanenitrile) from acetone. These examples show how reaction conditions can control the products of a reaction. While Le Chatelier's Principle controls ketal formation, pK_a controls the addition of HCN to a ketone or nitrile. Also, the rate of addition is strongly affected by the electron donating properties of the substituents. Aromatic aldehydes are less reactive than alkyl aldehydes and aromatic ketones less reactive still.

Typical reaction conditions use the sodium or potassium cyanide salt and acidify with sulfuric acid. The pK_a of hydrogen cyanide is 9.4. A useful application of the Henderson-Hasselbalch equation is that at pH 8.4, 10% of the hydrogen cyanide is ionized and at pH 7.4, 1% is ionized. Therefore, while acidification is required for the reaction, too low of a reaction pH would slow the reaction by decreasing the cyanide ion concentration.

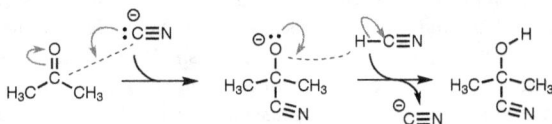

11. The cyanohydrin (2-hydroxy-2-methylpropanenitrile) is reversed with base to form acetone. This exemplifies an acid-base relationship in product formation as indicated at the start of the chapter. Because hydrogen cyanide is a stronger acid (pK$_a$ 9.4) than the alcohol (pK$_a$ ca 18), formation of the alkoxide can lead to a weaker cyanide base and drive the reaction to the right.

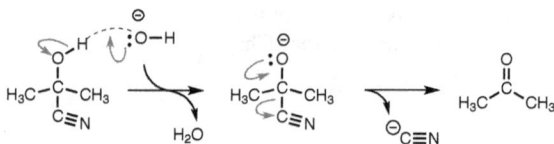

12. Addition of ethyl acetate enolate to 4-bromobenzaldehyde gives a benzyl alcohol. This example extends the nucleophiles that can be added. The formation of an enolate from an ester is discussed in Chapter 9.9

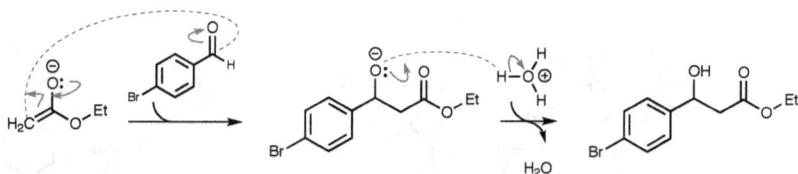

Reactions of Acyl Chlorides, Anhydrides, Esters, and Amides

Esters from Acid Chlorides or Anhydrides

13. A reaction of benzoyl chloride with ethoxide ion gives ethyl benzoate. As noted at the start of this chapter, it can be difficult to predict which intermediates are present in an addition reaction to a carbonyl group. This reaction could be carried out with pre-formed ethoxide or with a tertiary amine base. A tertiary amine base will not form a stable amide (Example 17), yet can form low concentrations of alkoxide and be an acid acceptor. This results in the mechanism of Example B. This mechanism is a good choice for this example because benzoyl chloride is only moderately reactive.

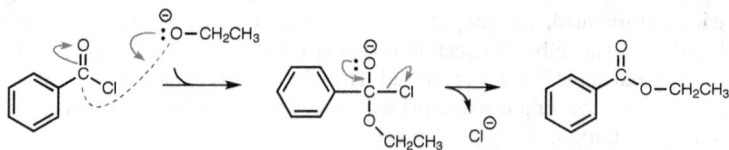

14. A direct reaction of ethanol with acetyl chloride gives ethyl acetate. This mechanism was particularly difficult to predict. Textbooks gave no guidance. I questioned whether a zwitterionic intermediate might form under the increasingly acidic conditions. Since HCl is released during the reaction, could the mechanism become acid catalyzed as in Example A? However, initially, the reaction must occur before an appreciable amount

of acid can be present, therefore it seemed plausible to form the zwitterion. I rationalized its formation by imagining the effect that an oxonium ion and a chloride would have on the electrons of the alkoxide. It seemed reasonable that these substituents would indeed stabilize the zwitterionic intermediate. From this intermediate, chloride is the weakest base and should be the first group to leave. The next intermediate is a protonated ester, a stronger acid than the preceding intermediate, and readily loses its proton to the alcohol solvent.

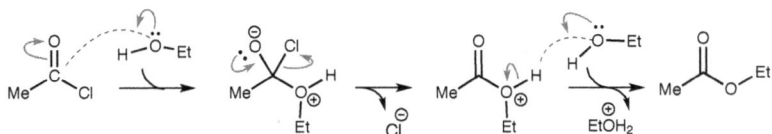

15. A pyridine catalyzed acylation with benzoyl chloride gives ethyl benzoate. Since pyridine is the stronger base, it can add in the first step. Then, chloride, as a weaker base then pyridine, is the best leaving group from the tetrahedral intermediate. Similarly, in the next intermediate, ethanol is a weaker base than pyridine, therefore it must be deprotonated to make pyridine the weakest base. Otherwise the addition of ethanol will reverse to the prior intermediate. The loss of pyridine from the tetrahedral intermediate gives the final product. This show the catalytic role of pyridine in an acylation reaction.

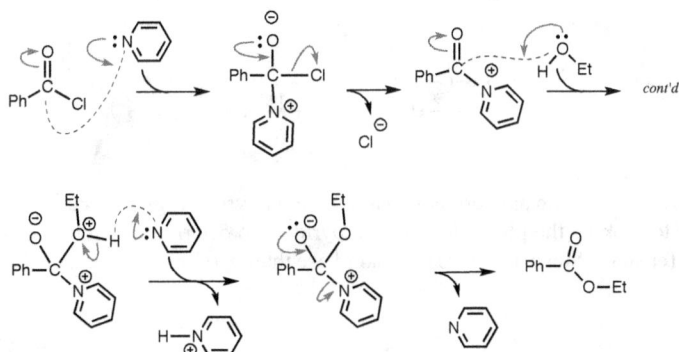

Using a pyridine catalyst is an appropriate option for an aroyl chloride as they are not as reactive as an acyl chloride. Also, pyridine is both a catalyst and base. An even better catalyst is 4-(N,N-dimethylamino)pyridine (DMAP) as it is more nucleophilic than pyridine, yet is still a good leaving group.

16. A reaction of acetic anhydride with ethanol catalyzed by sulfuric acid gives ethyl acetate. Anhydrides are less reactive than acid chlorides, therefore acid or base catalysts are commonly used to accelerate the reactions. I anticipate an intramolecular

proton transfer will occur as intramolecular proton transfers are entropically favored. If a base were used, a reaction mechanism similar to Example B would prevail.

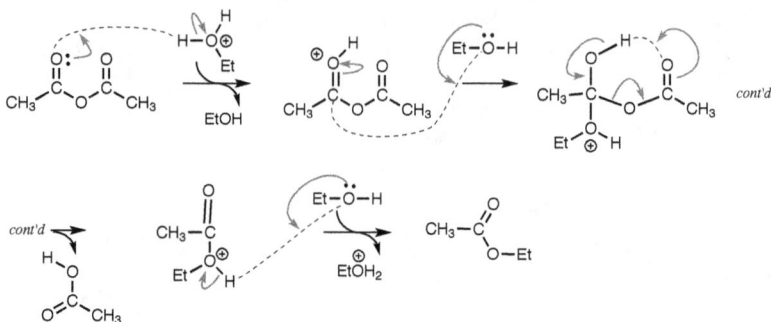

Amides from Acid Chlorides or Anhydrides

17. A reaction of benzoyl chloride with ethylamine gives *N*-ethylbenzamide. This example can be compared with Example 14. Because amines are more nucleophilic, they can react uncatalyzed with moderately reactive acyl groups. Consequently, these reactions can even be performed in protic solvents like ethanol or water. While these solvents can react with the acid chloride or anhydride, they do not react as fast as an amine nucleophile. This example could have been carried out in water with sodium hydroxide (Schotten-Baumen reaction).

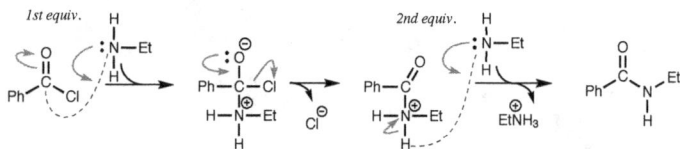

18. A reaction of acetic anhydride with aniline (and pyridine) gives acetanilide. You may need to look up the pK$_a$ of aniline and pyridine to determine the order of the proton transfer steps. Since aniline is a weaker base than pyridine, it should be deprotonated first.

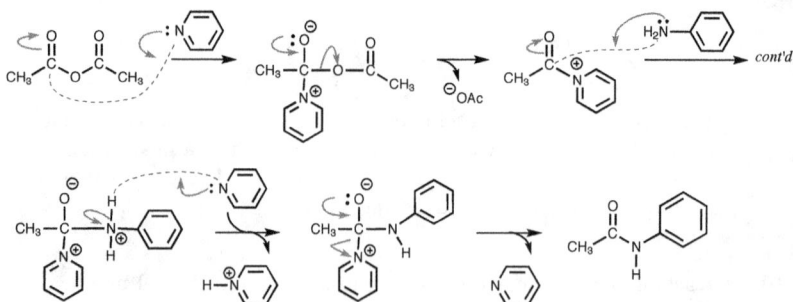

19. A reaction of acetic anhydride with aniline gives acetanilide.

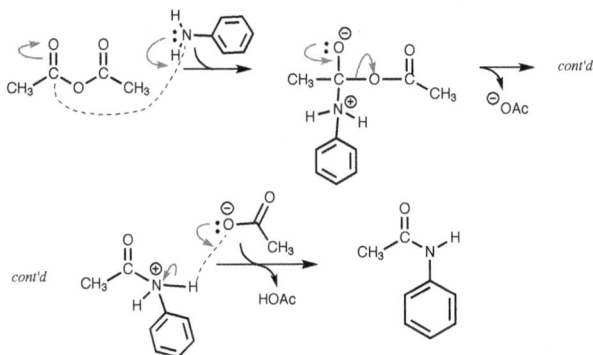

Ester from Acid with Mineral Acid Catalysis (Fischer Esterfication)

20. A reaction of benzoic acid and methanol with sulfuric acid or hydrogen chloride gives methyl benzoate. This is a carboxylic acid analog of the ketalization reaction in Example 7. This is an equilibrium reaction and the products are controlled by Le Chatelier's Principle. Although only a catalytic amount of acid is required, the acid also sequesters water to drive the equilibrium toward the ester.

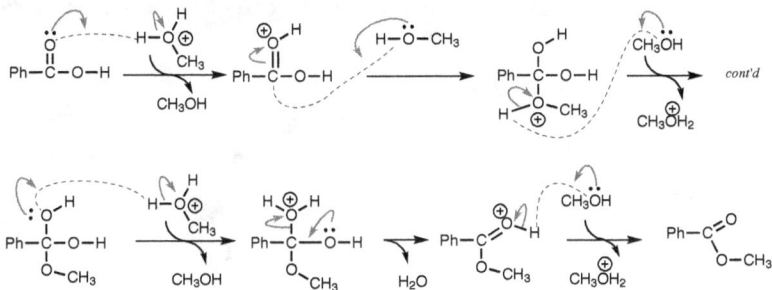

Acid Catalyzed Hydrolysis of an Ester

21. An acid catalyzed hydrolysis of methyl pentanoate (valerate) gives pentanoic (valeric) acid plus methanol. Just as Example 20 was an equilibrium favoring ester, the reaction can be reversed by addition of water to drive the equilibrium toward the acid.

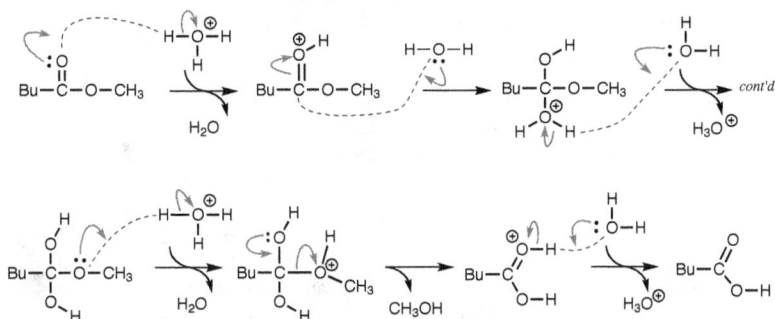

22. An acid catalyzed conversion of *t*-butyl pivalate to pivalic acid. This is a convenient alternative method to convert a tertiary butyl ester to a carboxylic acid. Because the trifluoroacetic acid and the reaction by-products are volatile, work-up is simple as well. Note the change in mechanism.

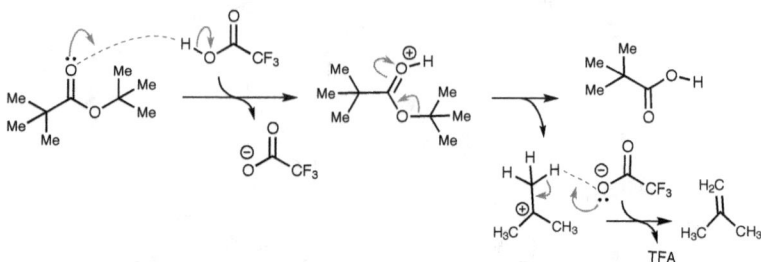

Base Hydrolysis of an Ester (Saponification)

23. Base hydrolysis of octyl isobutyrate gives octanol and isobutyric acid. Step 1 is the treatment with base. Mild reaction conditions* can be used with bases such as potassium carbonate† (potash) for making soap. The pK$_a$ of the tetrahedral intermediate alkoxide (see Example B, p. 64) is nearly the same as the starting material, therefore no barrier must be overcome in the formation of this intermediate. The intermediate will lose the weakest base, to reform the ester or form the acid. Both reactions occur, but the deprotonation of the carboxylic acid renders it irreversible.

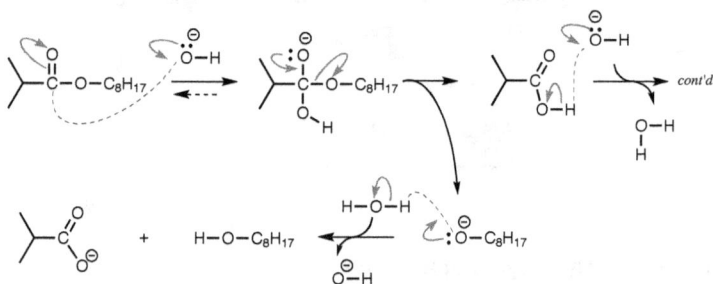

Step 2 is acidification of isobutyrate and isolation of isobutyric acid. If you wish to isolate the neutral compound, it should be extracted before acidification. Similarly, the alcohol should be removed before acidification to prevent contamination of the carboxylic acid.

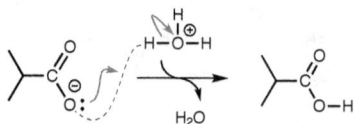

---

* With equimolar acid or base solutions, the difference between the pH and pK$_a$ of the intermediate for mechanisms in Examples A or B is smaller for basic than acidic reactions. Therefore, the reaction rate for a basic reaction is faster than for an acid catalyzed reaction.

† The pH of carbonate is circa 11 and generates enough hydroxide to react at the carbonyl group.

Hydrolysis of an Amide

24. Base hydrolysis of *N*-butylacetamide gives *n*-butylamine and acetic acid. Because nitrogen is much more basic than oxygen, intermediate **1** must be converted to the dianion **2** to expel a nitrogen atom under basic conditions. Water is the weakest base, therefore HO⁻ is the weakest base from intermediate **1**. In dianion **2**, the amide anion is the weakest base. The alternates, methyl anion or an oxygen dianion, are not weaker bases. The pK$_a$'s of CH$_4$, BuNH$_2$(or NH$_3$), and H$_2$O are 50, 36, and 15.7, respectively. The weakest base from intermediate **2** is BuNH⁻, the conjugate base of BuNH$_2$. Accordingly, basic hydrolysis of an amide requires higher concentrations of base and more vigorous condition.

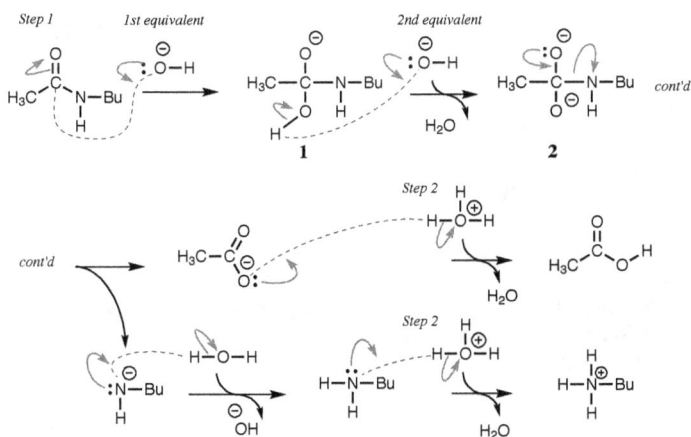

25. Acid catalyzed hydrolysis of *N,N*-dimethylacetamide gives acetic acid and dimethylammonium chloride. A nitrogen reduces the pK$_a$ of an amide, therefore the rate of an acid catalyzed hydrolysis is expected to be much faster‡ than a basic hydrolysis.

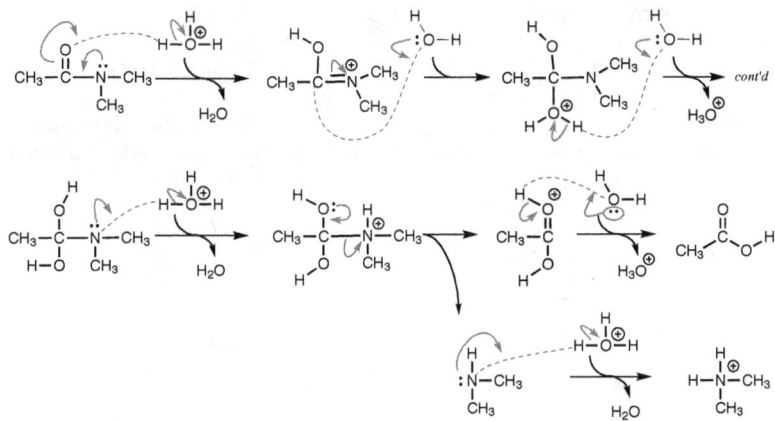

---

‡ With equimolar acid or base solutions, the difference between the pH and pK$_a$ of the intermediated for mechanisms in Examples A or B is smaller for acidic than basic reactions. Therefore, the reaction rate for an acid catalyzed reaction is faster than for a basic reaction.

Reactions of Esters

26. Addition of methyl magnesium bromide to ethyl benzoate gives 2-phenyl-2-propanol. This reaction is analogous to the previous additions to carbonyl compounds, however, the loss of ethoxide from the intermediate results in the formation of a new and more reactive carbonyl compound. Because the starting ester is less reactive than the intermediate methyl ketone, if less than 1 equivalent of Grignard reagent is used, then unreacted ester will be found as a by-product.

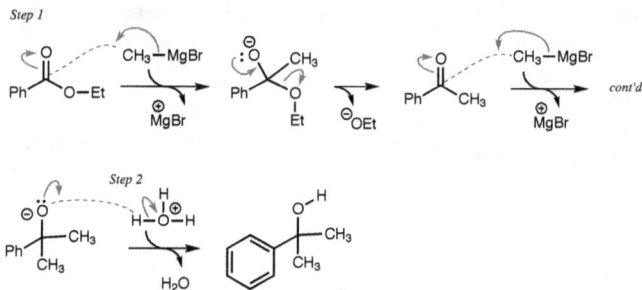

Step 1

Step 2

27. A reaction of 4-nitrophenyl propionate with ethyl amine gives *N*-ethyl propionamide. Esters are slow to react with an amine. Electron withdrawing groups attached to the ester, such as a trifluoroacetate, increase the rate of reaction. In this example, a *p*-nitrophenyl ester is used to make the oxygen a weaker base and a better leaving group.

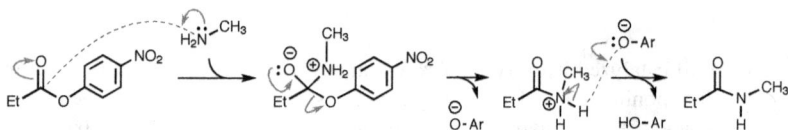

Reactions of Nitriles

28. An acid catalyzed hydrolysis of benzonitrile gives benzoic acid and ammonium chloride. Because acid catalyzed hydrolysis of a nitrile produces an amide intermediate that is more reactive than the nitrile, it is difficult to isolate this intermediate. It is possible to stop an acid catalyzed nitrile hydrolysis at the amide stage by limiting the amount of water.

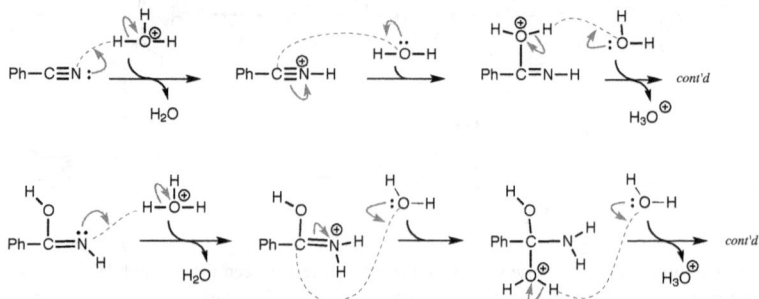

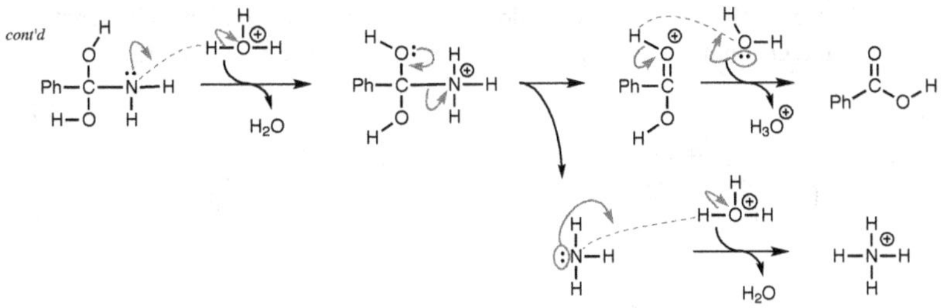

cont'd

Because the pK_a of the conjugate acid of an amide is greater than that of a nitrile, any amide that forms, is protonated more easily than the starting nitrile. As a result, the nitrile will be converted completely to the carboxylic acid. In this respect, the reaction is similar to a Grignard reagent adding to an ester where the intermediate is more reactive (basic) than the starting material.

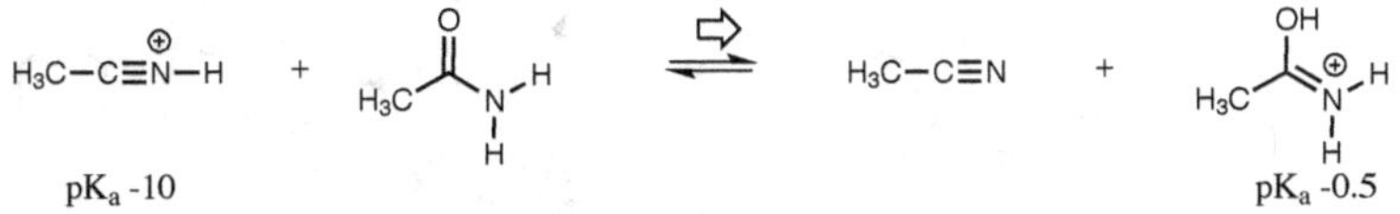

pK_a -10

pK_a -0.5

29. A base catalyzed hydrolysis of a nitrile can convert cyclopentanecarbonitrile to cyclo-pentanecarboxamide. Under basic conditions, an hydrolysis can stop at the amide stage. Because a dianion intermediate is required for a base hydrolysis of an amide, the reaction conditions must be more vigorous or an incomplete hydrolysis can result.

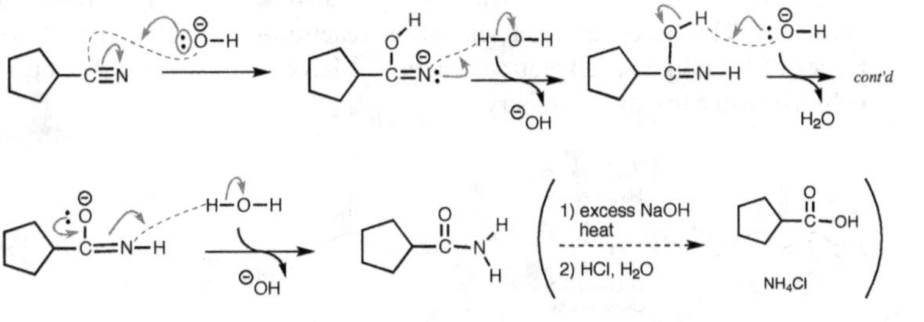

30. The addition of phenyllithium to a nitrile gives, after hydrolysis, cyclohexyl phenyl ketone. In step one of the addition to the nitrile, a Grignard reagent can only add one equivalent of Grignard reagent. The intermediate will resist increasing the negative

charge on the nitrogen from addition of a second equivalent. The reaction of a nitrile with a Grignard reagent results in a ketone because Example 30.

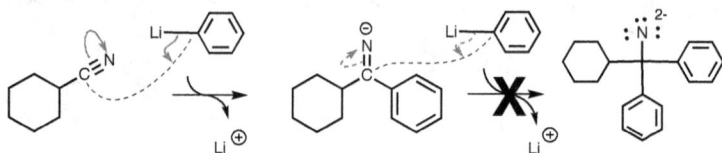

Step 2, hydrolysis of the imine.

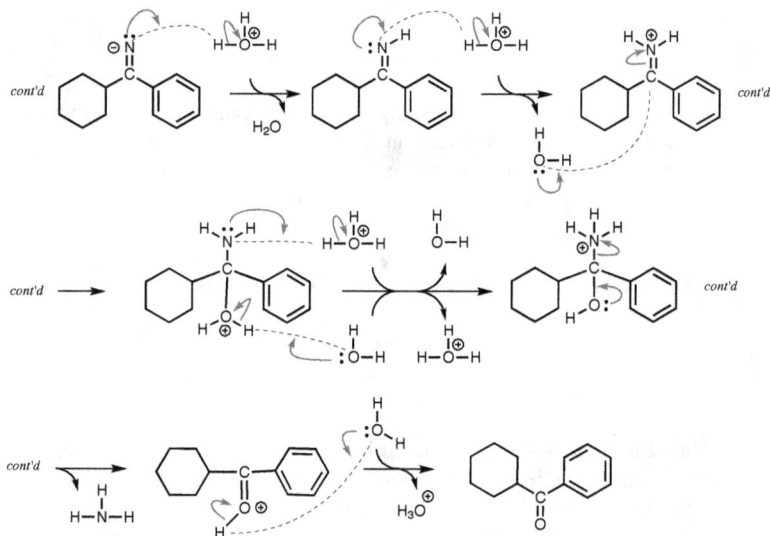

Miscellaneous, Ester Formation with Diazomethane

31. A reaction of diazomethane with a carboxylic acid forms methyl 2-butenoate (methyl crotonate). This is an unusual alkylation reaction of a carboxylate anion because it doesn't look like an alkylation reaction. I have placed it here because it is often included with carbonyl chemistry.

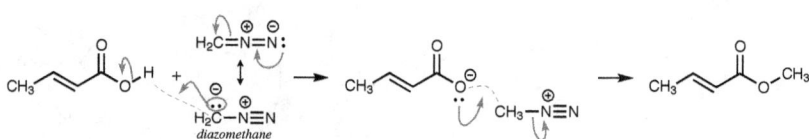

9 — Reactions of Enols and Enolates

Aldol Reaction

1. Stage one of a base catalyzed aldol condensation of butanal (butyraldehyde). This is a self-condensation reaction and there may be some confusion as to what species might be present. The reactants are in equilibrium, therefore an enolate and aldehyde are both present and are able to react with one another.

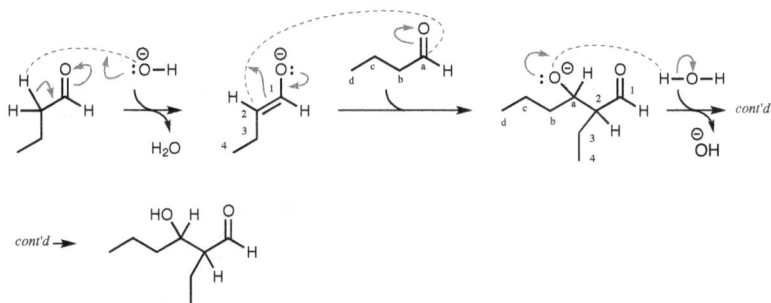

Stage two of a base catalyzed aldol reaction, the dehydration step. The dehydration step is frequently accomplished by heat or strong base (or strong acid). I infer from heating or using excess base to expect a dehydration reaction to occur. Similarly, low temperatures or a very low base concentration are often a signal that the dehydration will not occur, see Example 2. A base catalyzed dehydration is an E1cb reaction, see Chapter 4.

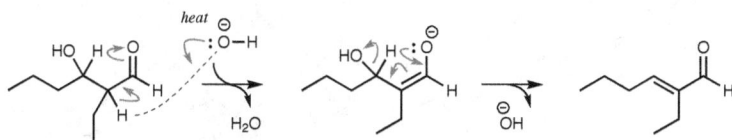

2. In a directed aldol condensation, in step 1 the enolate of 2-methylcyclohexanone is prepared with the non-nucleophilic base lithium diisopropylamide (LDA). In this case, it is the kinetic product. The aldol reaction occurs with propanal in step 2 and then the reaction is neutralized during work up.

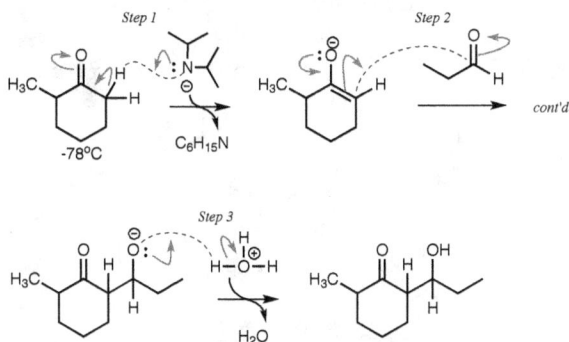

For a kinetic deprotonation of 2-methylcyclohexanone, you may choose a) the kinetic deprotonation removes the less sterically hindered hydrogen or b) deprotonation occurs faster on the more acidic hydrogens. Since carbon is a better electron donor, substitution will decrease the acidity at the more substituted carbon (see Chapter 1). However, the kinetic product is not the most stable enolate. The more substituted is the more stable. If an excess of ketone is present, then a protonation-deprotonation process results in the formation of the thermodynamic enolate.

This addition product has three chirality centers. While there is a major product that results, I have not indicated its stereochemistry.

3. A base catalyzed mixed or crossed aldol condensation of acetone and benzaldehyde is similar to the directed aldol of Example 2. It differs because only one component of the mixture can enolize. Therefore, if the ketone is added slowly to the aldehyde-hydroxide mixture, the enolate, once formed, can react with the more abundant aldehyde.

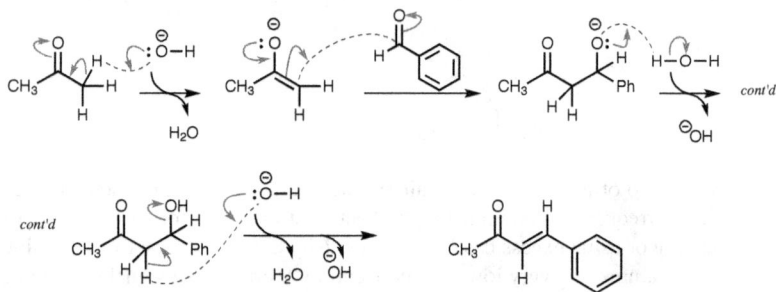

4. A Mannich reaction between 2-propanone, diethylamine, and formaldehyde gives 4-(diethylamino)butan-2-one. This reaction is like an acid catalyzed aldol condensation.

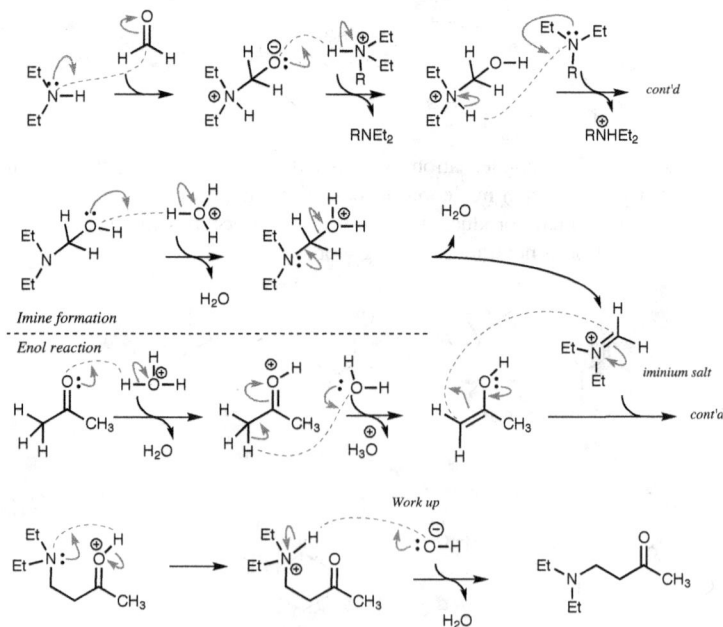

However, the Mannich reaction contains an extra initial imine formation step (see Chapter 8.9). Then protonation of the ketone and deprotonation results in the enol. The enol adds to the iminium salt in a manner similar to the aldol condensation.

I used a Mannich reaction as a substitute for an acid catalyzed aldol condensation. Although acid catalyzed aldol condensations have been performed, good examples of high yielding reactions are difficult to find. This mechanism has been generally drawn as an acid catalyzed enolization (and satisfies my objectives), but other mechanisms are possible.

In an acid catalyzed aldol condensation, first the enol of a ketone must form. The deprotonaton step will depend on the concentration of protonated ketone, a step that is not kinetically favored. Secondly, although protonation does increase the C-H acidity, the O-H acid remains stronger. Finally, addition of water would compete with deprotonation, and the pK_a indicates addition of water should be favored. In summary, these are reasons to expect enol concentrations to be lower in an acid catalyzed than base catalyzed reactions and water may slow the reactions further. Successful acid catalyzed aldol reactions are vigorously acidic and are accompanied by the removal of water.

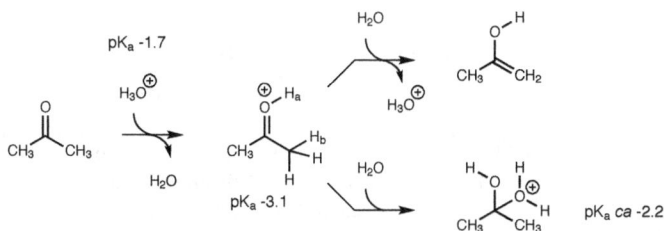

If one compares a base catalyzed enolization, this reaction should produce a larger concentration of enolate. The reaction of a ketone with hydroxide occurs from the unreacted ketone, therefore the rate of addition or enolization should be higher. Secondly, although addition will compete with enolization, the pK_as are similar. Therefore, no gradient should exist that will greatly favor addition.

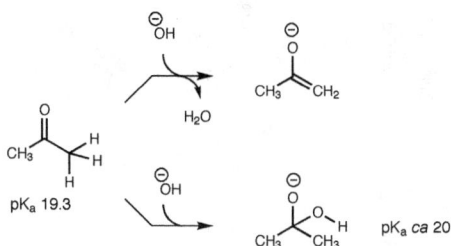

Returning to the Mannich reaction. The conditions for the reaction are mildly acidic aqueous conditions. Those conditions are favorable for imine formation step (*ca* pH 4). If the pH of the reaction were too low, then the amine concentrations would be shifted to the ammonium salt. The addition to the carbonyl group would be slowed or stopped completely.

The Mannich reaction is generally conducted with the amine hydrochloride salt plus a catalytic amount of additional acid. Because the Mannich reaction is a one pot reaction with all of the reagents present at the same time, the conditions favorable for the imine formation must also be compatible with an enol formation step. The acid catalyzed enolization step seems unlikely. A low pH may increase carbonyl protonation but at the expense of the presence of free amine.

I suggest an alternate mechanism be considered. Some free amine adds to the carbonyl group and forms some enamine, see Example 17. The enamine adds to the iminium salt, the new iminium salt hydrolyzes back to the ketone and this gives the product.

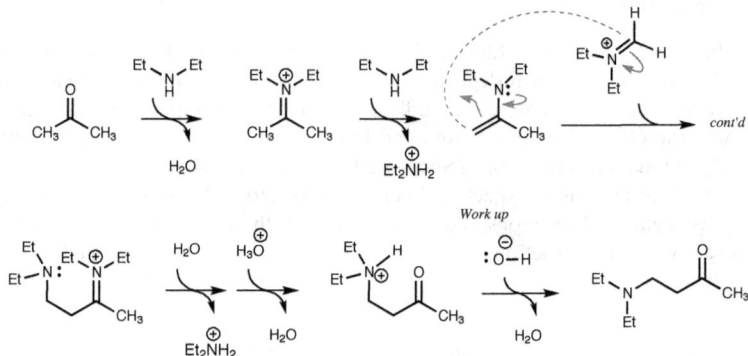

Claisen Condensation

5. An ethoxide catalyzed Claisen condensation of ethyl acetate gives ethyl acetoacetate (ethyl 3-oxobutanoate). This is another self-condensation reaction and it requires an excess of sodium ethoxide. The representation of reactant prefixes causes some confusion. While two moles of reactant form one mole of product, organic chemist generally use prefixes to indicate ratios of starting reactants. In a polymerization reaction, chemists don't use a prefix of n to indicate a large and undetermined number of moles are to combine. Similarly, a prefix may not be present in a self-condensation reaction. In the aldol condensation, only a catalytic amount of hydroxide ion is needed to catalyze this reaction. A prefix can incorrectly imply that one mole of hydroxide is used.

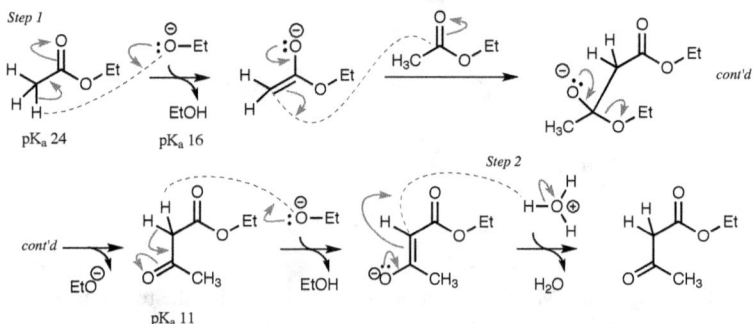

The pK$_a$ of ethyl acetate and ethanol results in an equilibrium favoring the ethoxide over the enolate anion. Therefore, the concentration of the enolate will be low.

The Claisen condensation product also has an acidic hydrogen. The pK$_a$ is lower than ethanol or ethyl acetate, therefore in the presence of ethoxide, the equilibrium is toward its enolate. This will consume ethoxide and stop the reaction before completion. The Claisen condensation is generally run with an excess of base to increase the concentration of enolate (because the equilibrium favors the neutral ester) and to overcome the consumption of base by the intermediate.

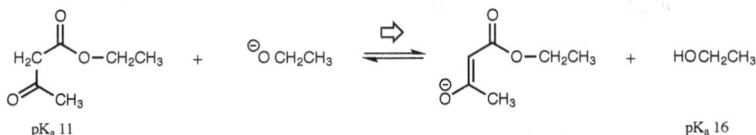

6. Ethoxide catalyzed crossed-Claisen condensation of cyclohexanone and ethyl formate. This crossed-Claisen condensation is similar to the crossed aldol condensation. The ethyl formate does not have any enolizable hydrogens, therefore it cannot form a nucleophile. The neutral product after work up prefers the enol form of the aldehyde. Aldehydes are easily enolized and the result is conjugated to the ketone.

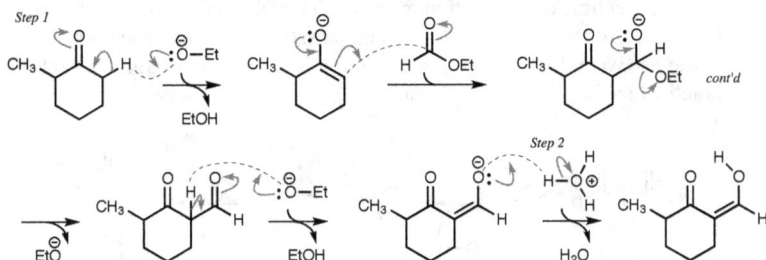

With the 2-methylcyclohexanone, two possible enolate and condensation products can and do form. However, the hindered product cannot enolize and therefore the reaction reverses to reform the enolate. The equilibrium favors formation of the weakest base, which is the product of this reaction. Example 11 is another reverse Claisen reaction.

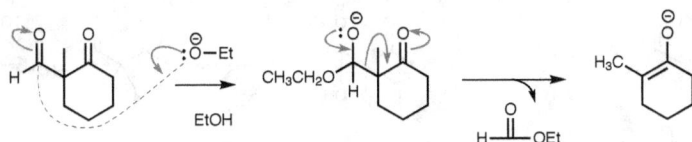

7. Ethoxide catalyzed crossed-Claisen condensation of ethyl acetate and ethyl benzoate. This crossed-Claisen can start with a catalytic amount of ethanol. The ethoxide generated can from an enolate plus ethanol. The ethanol can react with sodium hydride to generate more ethoxide to catalyze the reaction. This modification can be used to trap

protoic intermediates and increase enolate concentrations by making the enolization irreversible.

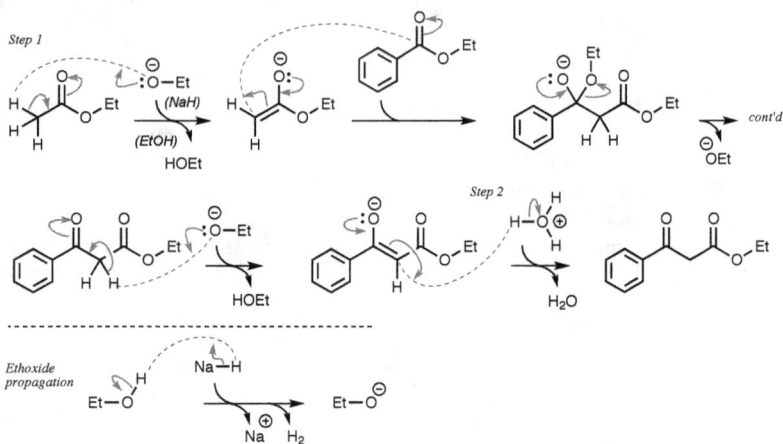

Step 1

Step 2

Ethoxide propagation

Acetoacetate Synthesis

8. Step 1, S_N2 alkylation of acetoacetate. This sequence of reactions accomplishes several goals. One, it reduces the pK_a of the nucleophile. Strongly basic nucleophiles are more susceptible to elimination reactions. Two, the stabilized enolate does not require low temperatures to avoid Claisen condensations that may accompany a direct alkylation as Example 9. Finally, this reaction is amenable to high concentrations and is practical for large scale preparations.

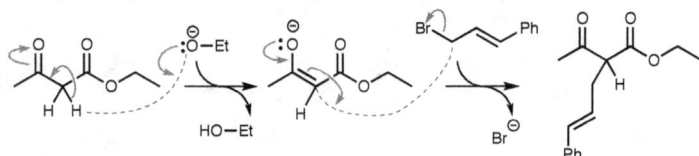

Step 2, sodium hydroxide hydrolysis of ester (saponification, Chapter 8.23). Alternately, an acid catalyzed hydrolysis and decarboxylation (step 3) can be combined into one step.

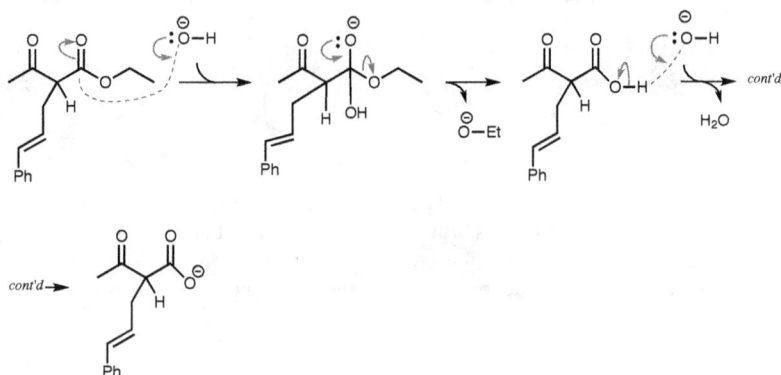

Step 3, acidification of carboxylate, decarboxylation, and tautomerization. This step utilizes a decarboxylation of a *beta*-keto acid. That grouping allows a proton transfer to form an enol in an intramolecular reaction. This process is entropically favorable and will occur upon heating. If an acid catalyzed hydrolysis is performed, the decarboxylation step will accompany the heating step used to accelerate the hydrolysis. If a malonic acid derivative is used, the same reaction takes place though the decarboxylation is less facile.

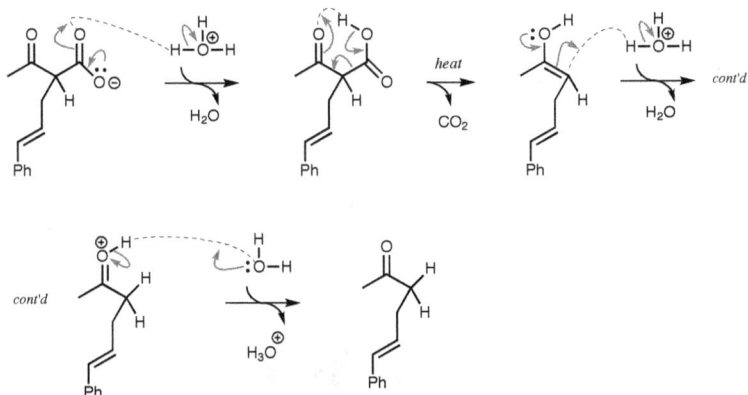

Enolate Alkylation Reactions

9. Enolization and alkylation of ethyl propionate with benzyl bromide. The advent of low temperature formation of enolates has enabled a direct alkylation of an ester.

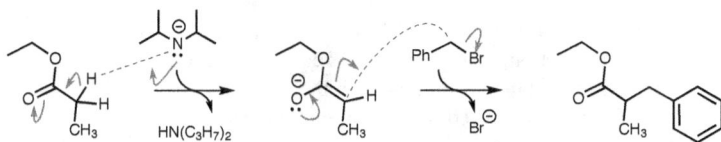

10. Sequential alkylation of a dianion of a *beta*-ketoester. The dianion can be easily rationalized with the principles of Chapter 1. I suggested acidity could be explained with a simple Coulombic model of proton-electron pair distances. Low pK_a acids should have greater proton-electron pair distances and weak acids smaller. The electrons of a low pK_a acid should be tightly held while those of higher pK_a acids should be weakly held. Consequently, the first alkylation reaction should occur on the high pK_a carbon

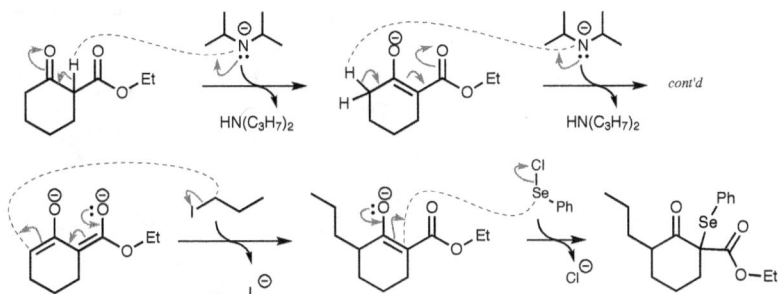

because those electrons should extend further from that carbon. The second reaction should occur on the more tightly held electrons. In this case, it forms the selenide, a *cis*-elimination precursor, see Chapter 4.19.

11. A retro-Claisen reaction from a *beta*-ketoester. This reaction should be noted for why Claisen condensations are not used with disubstituted esters. This example also shows that a slow irreversible reaction can overcome unfavorable pK_a conditions. The ester (enolate) is less acidic than ethanol, but if formed, the *beta*-ketoester does not reform.

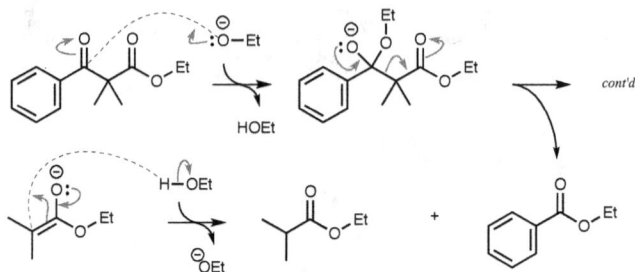

12. Enolization and alkylation of phenylacetonitrile with methyl iodide.

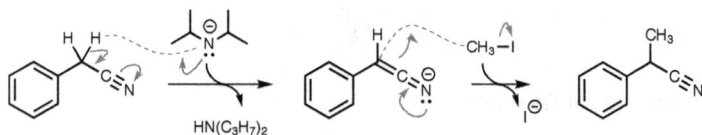

Halogenation of Carbonyl Compounds

13. A basic bromination of 3-methyl-2-butanone with sodium hydroxide and bromine give a bromoform reaction. This base catalyzed halogenation continues a theme started with the Claisen condensation in which the product is more acidic than the starting material. If a basic catalyzed monobromination reaction were attempted, the product would be 2/3 moles of recovered starting material and 1/3 mole a carboxylic acid and the trihalomethane. Since a tribromomethyl anion is more stable than a methyl anion (see p. 1B), it can form from cleavage of the carbonyl-hydroxide tetrahedral intermediate.

I also need to remind everyone that the addition to the carbonyl group that occurs in the last stage of step one, will have been occurring through the entire sequence of the reaction. It isn't a step that *only* occurs at the last step, although it will kinetically become faster as the halogens pull electrons away from the carbonyl group.

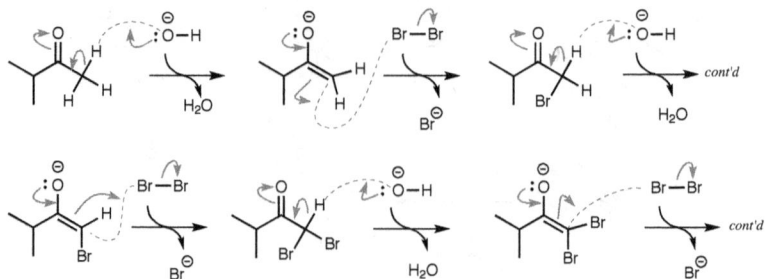

cont'd ⟶

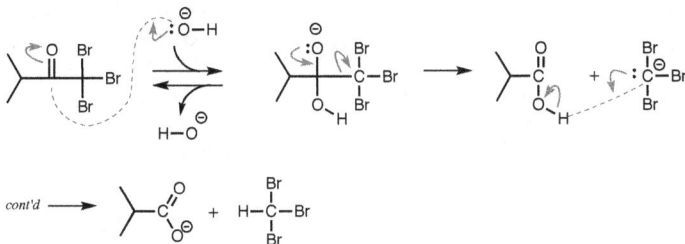

Step 2, acidification and isolation of isobutyric acid. If the acid is to isolated, extract the tribromomethane before acidification.

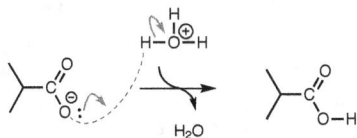

14. An acid catalyzed bromination of acetophenone gives α-bromoacetophenone. The rate-limiting step of the acid catalyzed bromination is the enolization of the ketone. In the acid catalyzed enolization, where does HBr come from? If bromine reacted with acetic acid, the products are HBr and CH₃COOBr. Also, the by-product of the bromination reaction is HBr. Furthermore, since bromine removes electron density from the carbonyl group, the starting material is more basic than the brominated product. Therefore, monobromination is more easily done under acidic conditions. The reaction rate is a function of the rate of enolization.

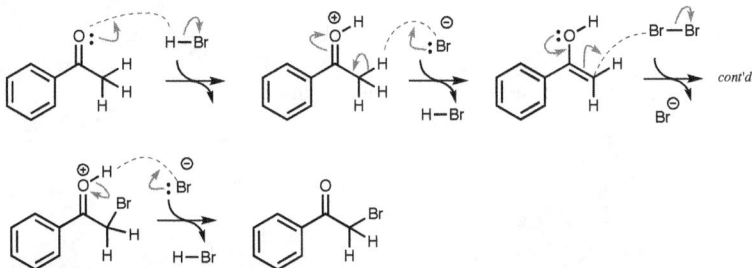

Michael or 1,4-Conjugate Addition Reaction

15. A Michael or 1,4-conjugate addition reaction of dimethylamine to methyl vinyl ketone.

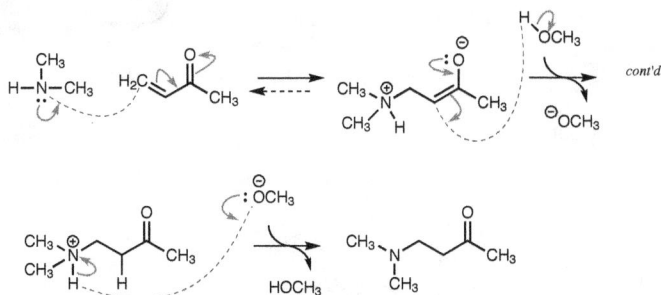

I rationalize conjugate addition reactions as the product of two competing modes, a fast addition to a carbonyl group (one pair of electrons) and a slower addition at the double bond (two pairs of electrons). If tetrahedral adduct **3** and then **4** were to form and the nucleophile was a weak base, then either could reverse to reform the starting materials.

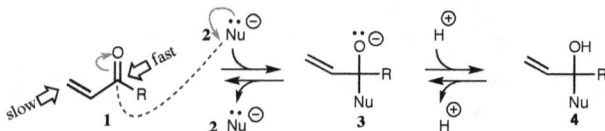

If a 1,4-addition reaction took place to form **5**, it too could reverse. However, if a protonation reaction also occurred to give **6**, it is less acidic and less able to the reform starting materials. If the nucleophile was a base of a weak acid (less than pK_a 20), it would not be basic enough to regenerate the enolate to reverse the reaction. If the nucleophile is the base of a very weak acid (pK_a greater than 20), then the tetrahedral intermediate **3** will be the most stable product because it will be the weakest base and that reaction would not reverse. The best nucleophiles for Michael addition or conjugation addition reactions are weak bases (less than an alkoxide). For the enone (Michael acceptor), if an alkyl groups is attached to the alkene, because they donate electrons, that will also slow conjugate addition reactions.

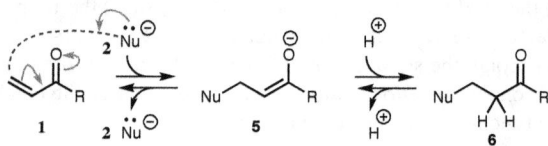

16. A Michael addition reaction of ethyl acetoacetate to cyclohexenone. This reaction only requires a catalytic amount of ethoxide. Therefore, only a portion of the final product would exist as the indicated anion and as a result, the acidification step could be skipped with little effect on the reaction yield.

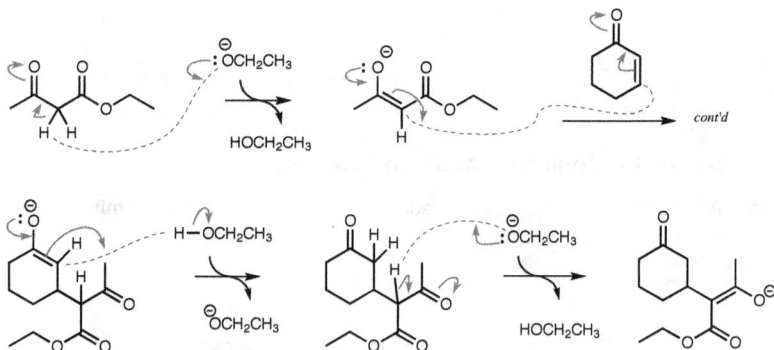

Step 2, neutralization of reaction mixture.

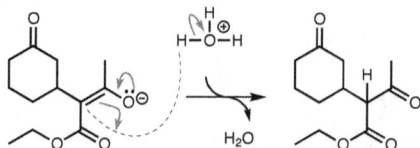

Enamine Alkylation

17. An enamine alkylation is a three step process. In Step 1, the morpholine enamine of cyclohexanone is formed. The mechanism of this reaction parallels other nucleophilic additions to carbonyl groups. A nitrogen atom is better able to add to a carbonyl group and thus does not need a catalyst. Catalysis is likely involved in the dehydration steps. It is similar to the optimal pH for oxime formation (*ca* 4). If the reaction pH were too low, the amine would be complexed by the acid. If the pH were high, the oxygen must be lost as hydroxide. In between, the oxygen can be lost as water. You should also note that because only a catalytic amount of acid is present, the predominant acid in the reaction will be an ammonium salt. This leads to an interesting protonation reaction, the protonation of the oxygen in the bottom left structure by an ammonium salt. Certainly the pK_a of an ammonium acid is higher than an oxonium acid. However, due to the electron donation of the nitrogen atom, the electrons on the oxygen are now held less tightly than normal, enabling to the protonation reaction.

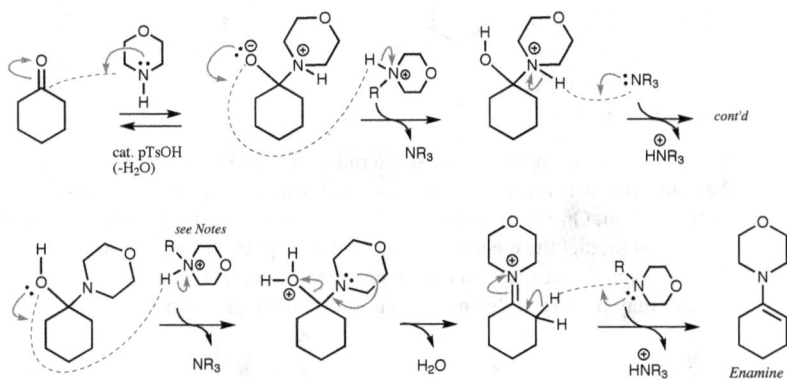

cat. pTsOH
(-H₂O)

cont'd

see Notes

Enamine

After water is lost, the product is an iminium salt, the nitrogen isostere of a carbonyl group. If R is H, then deprotonation occurs to give an imine and if R is not H, then the alkyl hydrogen is lost and the enamine results.

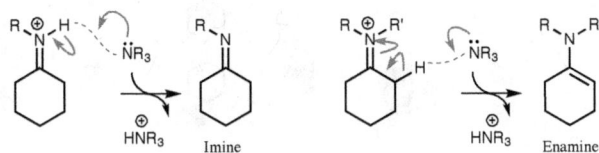

Imine

Enamine

Step 2, enamine alkylation with methyl iodide.

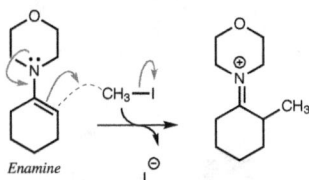

Enamine

Step 3, hydrolysis of iminium salt and isolation of substituted cyclohexanone.

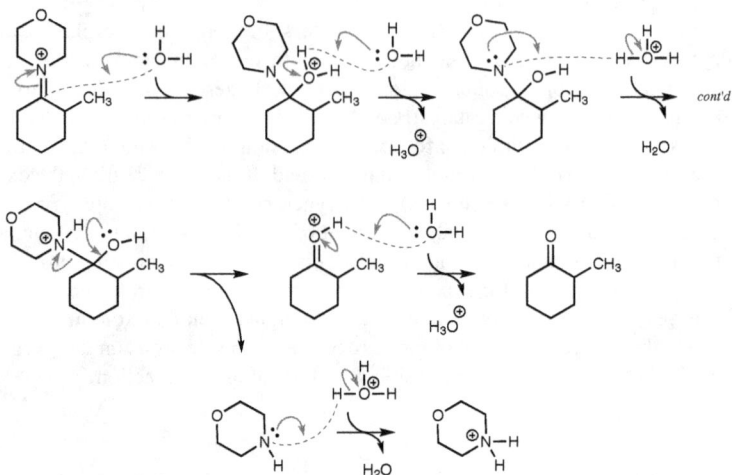

cont'd

Enamine Acylation

Why does an enamine react on carbon rather than the nitrogen atom? A key is to draw the resonance structure of the enamine. The resonance structure has a negative charge on the carbon. On which atom will the electrons extend the furthest? In which form would you predict the electrons are best able to react? However, many reactions may involve some direct reaction on the nitrogen atom. If it is an acylation reaction, that product may be reversible and leads to the same final product.

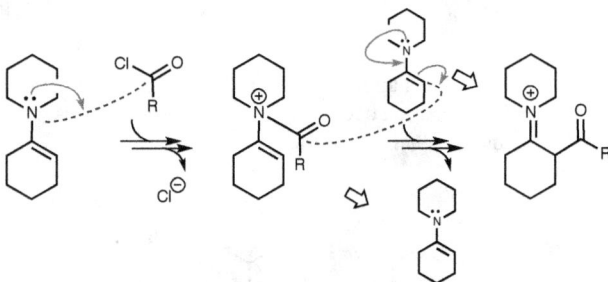

10 — Dehydration/Halogenation Agents

This chapter will deal with reagents that result in a net removal of water in which the oxygen atom becomes absorbed by the reagents. For example, one can perform an S_N2 nucleophilic substitution of an alcohol in the presence of HBr to give an alkyl bromide and water, see Chapter 3. In this chapter, we will see similar reactions, but water will not be formed as the leaving group.

1. A reaction of cyclohexanecarboxamide with acetic anhydride gives cyclohexanecarbonitrile. The dehydration of an amide with acetic anhydride is one of many reagents for this conversion. The mechanisms are fundamentally the same. They start with a complexation of the oxygen to make it a better leaving group. I like using acetic anhydride because the reaction mixture is simply heated and the by-products can be distilled off.

    The site of reaction is the oxygen atom rather than the nitrogen, see *Amide Regiochemistry*, p. 7.

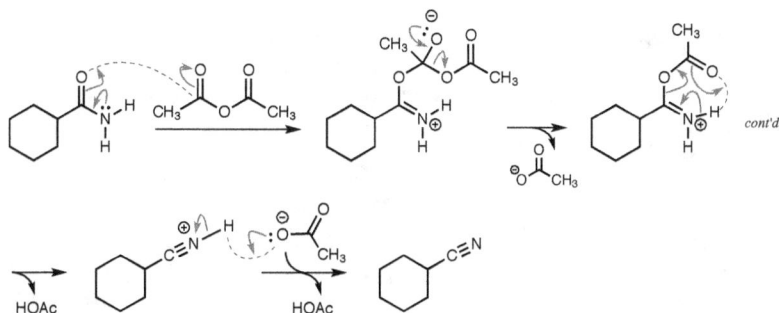

2. An S_N2 reaction of thionyl chloride with 1-butanol gives chlorobutane. This is one of two possible mechanisms for thionyl chloride reactions with alcohols. Pyridine is a base in the reaction, but it also prevent the loss of HCl. This increases the chloride ion concentration and the S_N2 substitution rate. In the absence of pyridine, then the chloride attack must be an intramolecular reaction from the same side, an S_Ni reaction. Substitution from the same side is thought to be an unfavorable reaction and as an alternate, it is thought that an ion pair mechanism may operate. With a chiral alcohol and pyridine, a chiral chloride of the opposite configuration results. Without pyridine, a chiral chloride of the same configuration results.

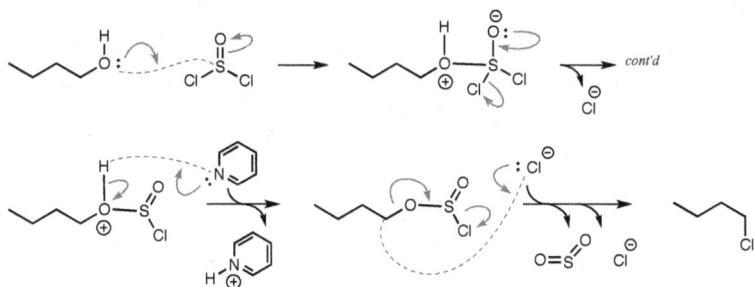

3. The reaction of 1-butanol with tosyl chloride and pyridine gives butyl tosylate. This example is similar to the thionyl chloride reaction. With toluenesulfonyl chloride, the product is a sulfonate ester rather than a chloride. Why doesn't the chloride ion by-product displace the tosylate ester? The chloride reaction is slower. If the reaction time is extended, chlorobutane will accompany the tosylate.

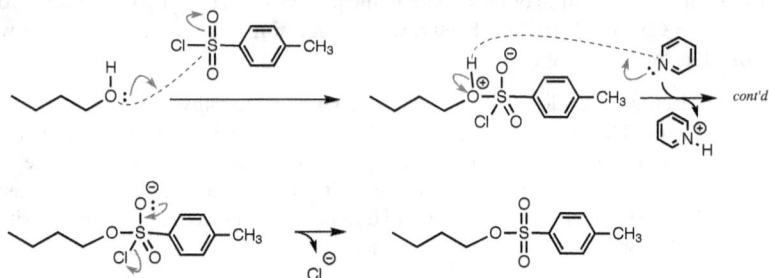

cont'd

Carboxylic Acid with Thionyl Chloride

4. The conversion of acetic acid to acetyl chloride with thionyl chloride is the classic reaction of thionyl chloride with a carboxylic acid. However, Example 5 is one of the best ways to make an acid chloride. It only requires a catalytic amount of DMF.

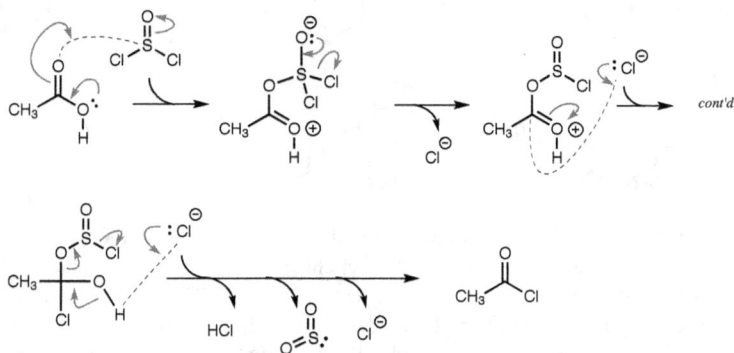

cont'd

5. An improved conversion of acetic acid to acetyl chloride with thionyl chloride. This is a one pot reaction with a catalytic amount of DMF. Part 1, reaction of thionyl chloride with DMF to form an iminium salt.

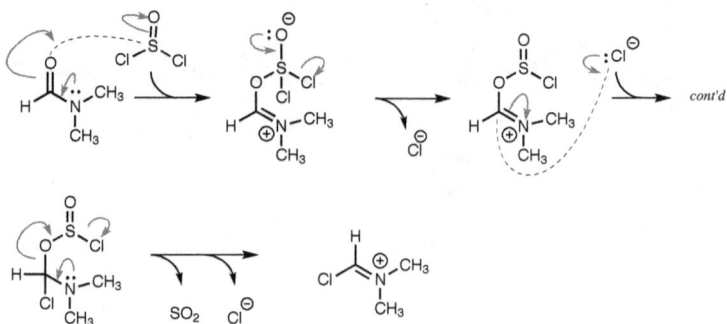

cont'd

94 • Chapter 10 — Dehydration/Halogenation Agents

Part 2, reaction of iminium salt with carboxylic acid. This forms acetyl chloride and reforms DMF to be recycled in Part 1.

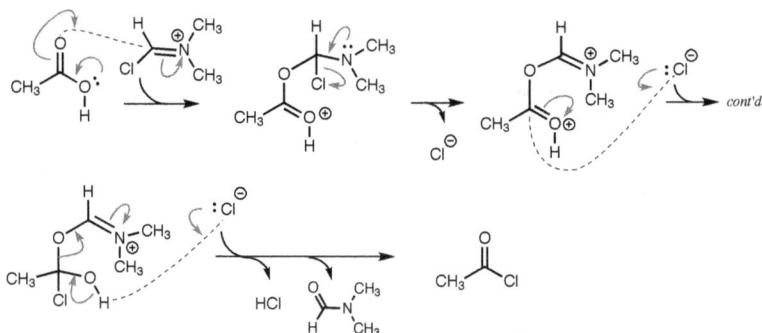

Halide from an Alcohol with a Phosphorus Reagent

6. Reaction of isobutyl alcohol with phosphorus tribromide gives isobutyl bromide.

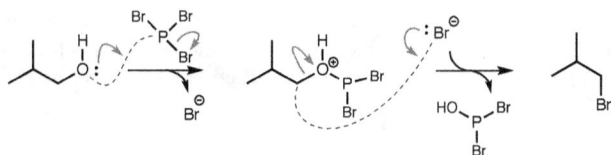

This example may be too simple. I have written what I think is a more likely mechanism(s) for this reaction below. You should note the boxed structures. A similar structure is present in the Arbusov reaction with a trialkyl phosphite (see Chapter 8.6, Step 1). Here it reacts with HBr gives an alkyl bromide and phosphorus acid. Therefore, formation of a phosphorus salt is a plausible mechanism.

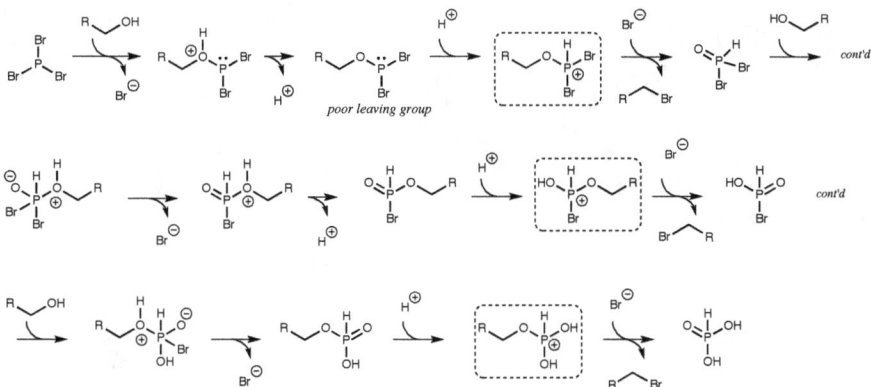

7. A reaction of triphenylphosphine, carbon tetrachloride and cyclopentanol gives chlorocyclopentane. This is a convenient and effective chlorination reaction. Carbon tetrachloride can be substituted with chlorine or N-chlorosuccinimide (NCS) for chlori-

nation reactions and with carbon tetrabromide, bromine, *N*-bromosuccinimide (NBS) and for bromination reactions.

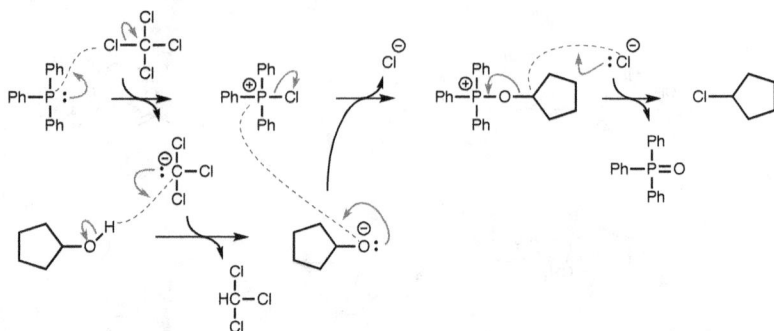

Ester from an Alcohol and Carboxylic Acid with a Phosphorus Reagent

8. The Mitsunobu reaction with triphenyl phosphine and diethyl azodicarboxylate is used to prepare an ester from an acid and an alcohol. The Mitsunobu reaction extends the phosphorus chemistry in an interesting way. Two protons are absorbed by the azo-compound and the oxygen becomes attached to the phosphorus for a net loss of water. The displacement is an S_N2 substitution reaction resulting in inversion of the alcohol.

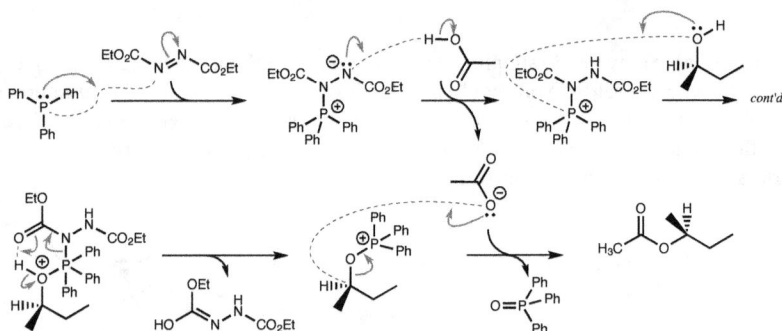

11 — Reduction Reactions

Sodium Borohydride Reductions

1. The sodium borohydride reduction of 2-methylpropanal (isobutyraldehyde) gives 2-methyl-1-propanol (isobutyl alcohol). In borohydride reductions, the solvent, methanol, is part of the reaction cycle. Each reduction generates a borane product which acts as a Lewis acid. Reaction with the solvent regenerates a borohydride reductant in which the hydride atoms are replaced with methoxide or an alkoxide. Thus, all of the hydride hydrogens can be used.

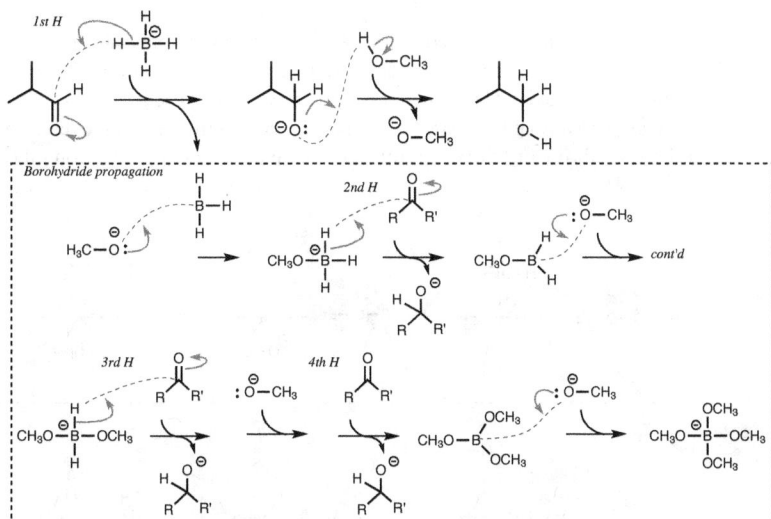

2. Sodium borohydride reduces cyclopentanone to cyclopentanol.

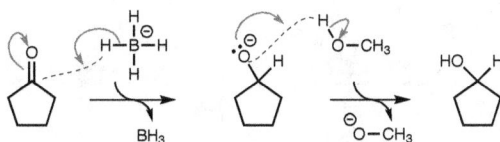

3. Sodium borohydride reduces ethyl 4-oxocyclohexanecarboxylate to ethyl 4-hydroxycyclohexanecarboxylate. Because esters are less reactive than aldehydes or ketones, a selective reduction can be done. Sodium borohydride does slowly react with esters, so a very reactive might also be reduced.

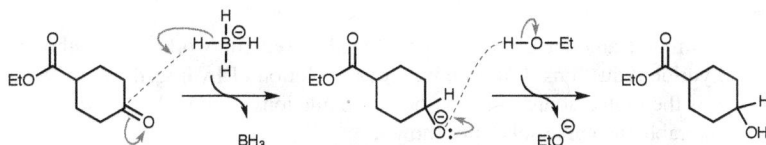

Lithium Aluminum Hydride Reductions

4. Lithium aluminum hydride (LAH) reduces acetophenone to 1-phenylethanol.

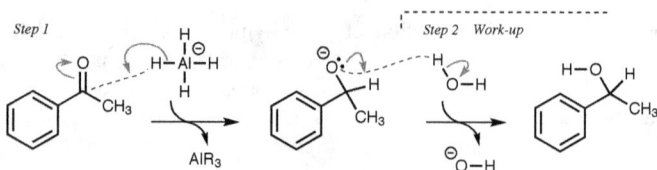

Step 1   *Step 2 Work-up*

5. Lithium aluminum hydride reduction of ethyl butanoate (butyrate) gives ethanol and butanol. In lithium aluminum hydride reductions, a cycle similar to a borohydride reduction occurs except that since an aprotic solvent is used, the trivalent aluminum must react with any of the Lewis base species present in a reaction cycle. The R-group could be an ethoxy or butoxy group as both are present in the reaction mixture.

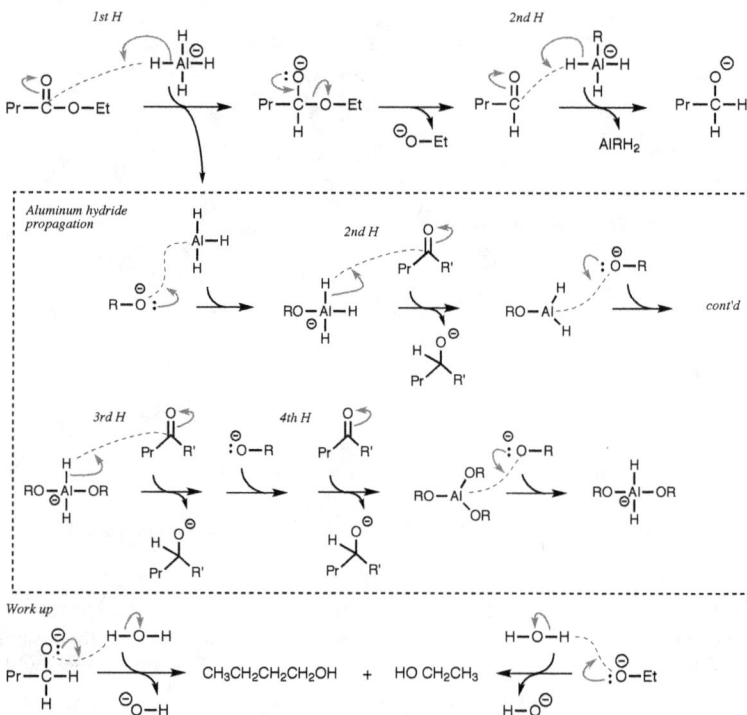

A limited amount of water is frequently used for working up lithium aluminum hydride reductions. Caution, vigorous evolution of hydrogen gas may ensue. While it is the proton source shown, the hydroxide ions react with the aluminium to form a filterable insoluble solid, not shown.

6. Lithium aluminum hydride reduces a cyclic amide (1-methylpyrrolidin-2-one) to give a cyclic amine (1-methylpyrrolidine). An amide reduction differs from other LAH reductions because an oxygen atom must be removed to get to the final amine product. In the reduction, an R-group can form from the alkoxy intermediate for regeneration of aluminum (IV) reductant. Also, the preferred form in which oxygen can be expelled is after donation of a hydride. I think this might be the most difficult step in the reaction and reductions which involve loss of a complexed oxygen are more sluggish than other LAH reductions. For example, references often show the use of an excess of LiAlH$_4$ to give a successful reduction of an amide.

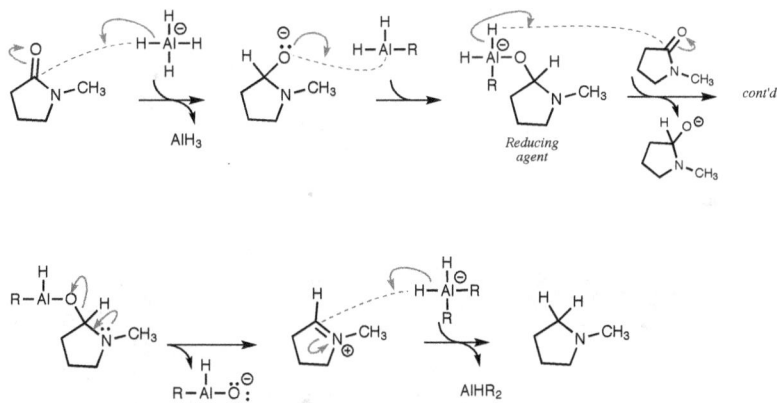

The preferred leaving group for lithium aluminum hydride reductions is shown with its resonance structure.

7. A Lithium aluminum hydride reduction of propionyl anilide gives *N*-propylaniline. Notice the lithium aluminum hydride reduction begins with an abstraction of a hydrogen. I needed an example to remind students of this reaction with an amide and carboxylic acid. As lithium aluminum hydride reacts at the start of a reaction, it reacts similarly with water (or alcohol) and therefore the entire reduction cycle must be carried out in the absence of moisture. Reductions are carried out in aprotic solvents such as ether, tetrahydrofuran, or dimethoxyethane. The release of hydrogen is also evident from unreacted aluminum hydrides during work-up. Lithium aluminum hydride will react violently with water or low molecular weight alcohols.

Step 1, reduction.

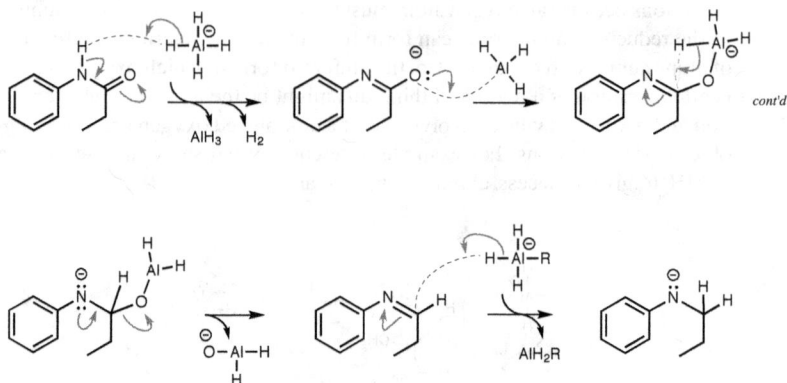

Step 2, work up.

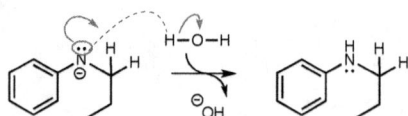

8. A lithium aluminum hydride reduction of phenylacetic acid gives 2-phenylethanol.

Step 1, reduction.

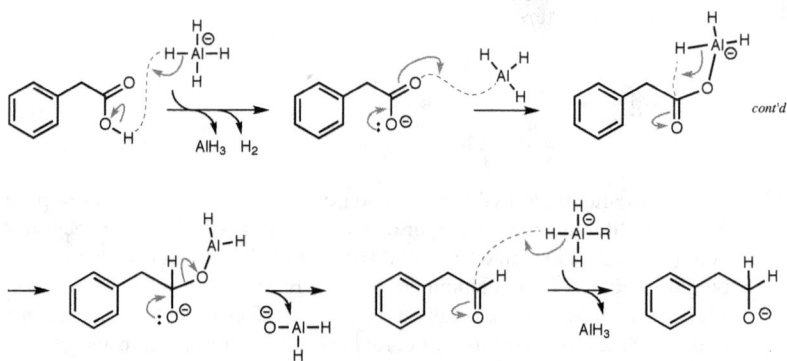

Step 2, work up.

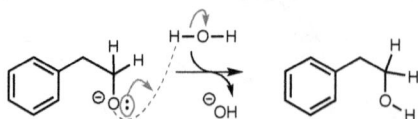

Reductive Amination

9. A reductive amination is a two-step process. The reactions are compatible with one another and are frequently carried out in a single flask. The imine, as it forms, can be reduced. In Step 1, an imine is formed from benzaldehyde and ethylamine.

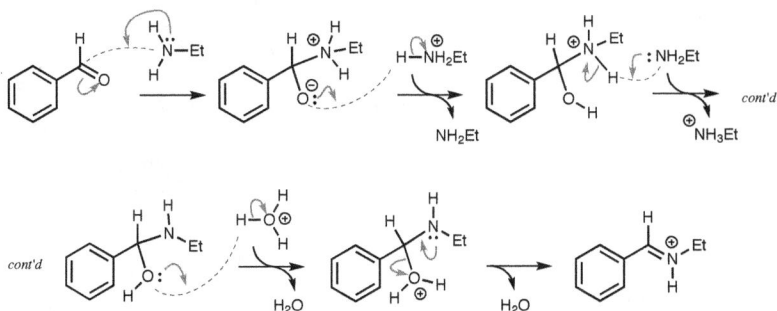

Step 2, reduction. Triacetoxyborohydride is one of several reagents that one can use. Others are sodium cyano-borohydride (NaBH₃CN) or hydrogen with a palladium or nickel catalyst.

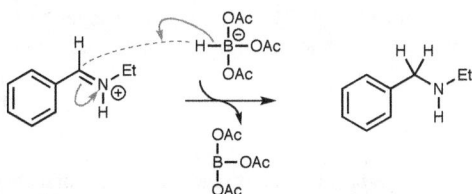

Diisobutylaluminum Hydride Reduction of an Ester

10. Step 1, addition of DIBAH or DIBAL(H) (diisobutylaluminum hydride) to ethyl isobutyrate. Reduction gives isobutyraldehyde (2-methylpropanal).

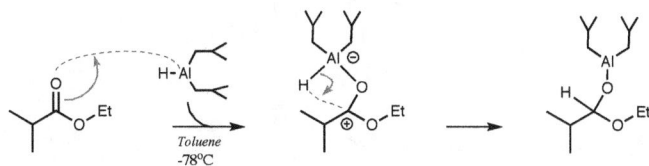

Step 2, work-up

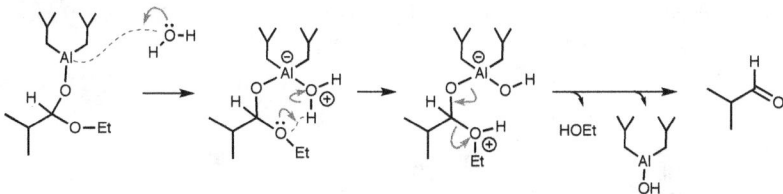

Reduction of Alkyne with Sodium and Ammonia

11. A sodium and ammonia reduction of 2-butyne gives *trans*-2-butene. You must use a different arrow for single electron transfers than normal. While most reactions involve reactions with two electrons, this reaction has two single electron transfer steps. They are noted with the single barbed arrow.

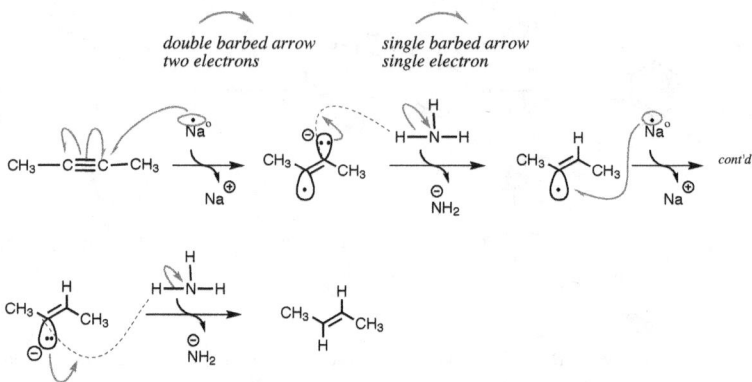

double barbed arrow
two electrons

single barbed arrow
single electron

The intermediates of this reaction are subject to an acid-base reaction. The pK$_a$ of the alkenyl anion is 44 and NH$_3$ is 36. The alkenyl anion is the stronger base, therefore it will abstract a proton from ammonia. The equilibrium favors formation of sodium amide as the conjugate base.

Wolff Kishner Reduction

12. A Wolff Kishner reduction is a one-pot, two step reaction. In the first reaction, the ketone reacts with hydrazine under basic conditions to form the hydrazide.

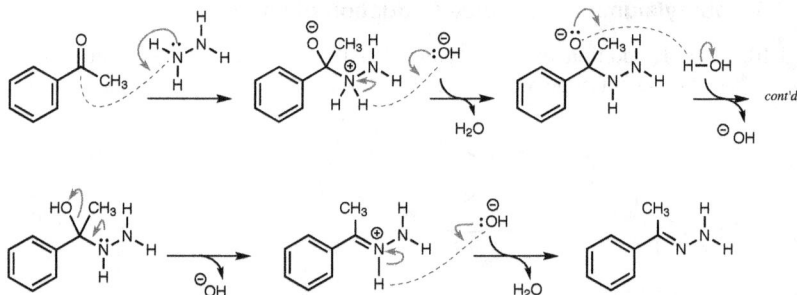

Reaction of the *in situ* formed hydrazide with potassium hydroxide to form a methylene and complete a reduction of the carbonyl group.

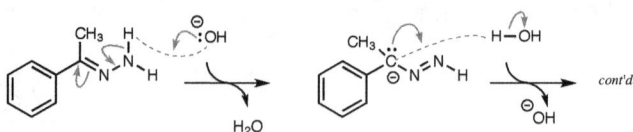

cont'd

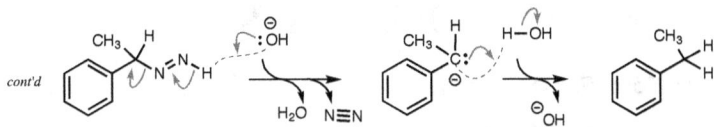

Catalytic Reduction of Nitrobenzene

13. Catalytic reduction of nitrobenzene to aniline. This isn't a mechanism. It shows that hydrogen can add across N-O single and double bonds. It is a very facile reduction. One can see that the oxygen atoms form water and the reduction takes three moles of hydrogen.

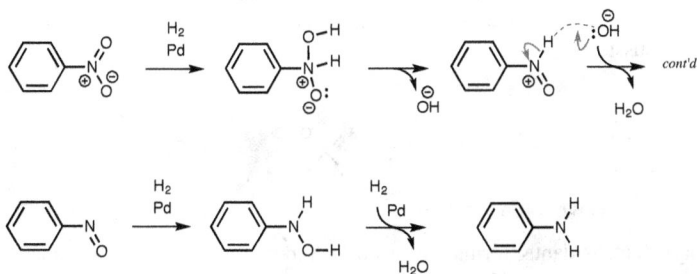

cont'd

12 — Oxidation Reactions

General Form For Oxidations

The key step in an oxidation reaction is to break an oxygen-oxidant bond (O-X) such that the X-group accepts the electrons, Equation 1. The group X can be a halogen, oxygen, or a metal, frequently Cr, Mn, Ag, and others.

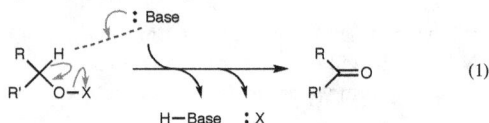

$$(1)$$

The first step to an O-X bond formation for monovalent oxidants (Br, Cl, Ag), is a simple displacement reaction, Equation 2. Those reactions are simple and generally don't need further discussion.

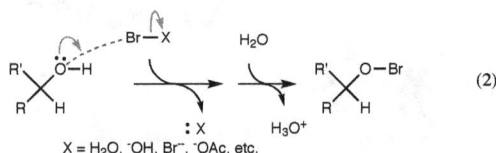

$$(2)$$

For polyvalent oxidants, formation of an O-X bond is lengthier. It is usually a series of proton transfer reactions. However, these proton transfer reactions often distract from the oxidation mechanism itself. Therefore, I am going to illustrate some principles of those reactions, some variations to consider, and an example or two. I will leave it to you to devise plausible steps to complete the reaction.

In the scheme below, four general routes to form an oxidant-ester are shown. The routes differ by the timing of proton transfer reactions, concurrent, zwitterionic, early protonation, or early deprotonation. Examples **A** and **B** are low and high pH mechanisms. Example C in an ionic route and one must determine whether these intermediates are consistent with the pH requirements of the reaction. Example **D** is electronically neutral, but places a higher entropic requirement on the reaction. Because these reactions precede the oxidation step, they are fast, not rate limiting, and their mechanisms are not known. Little literature data exists to guide us in determining their route. Therefore, the steps showing the formation of the esters of the oxidants are not described.

Example	Reaction
A	
B	
C	

D | R—Ö—H + (M with O double bonds) ⇌ [R-O...H...O / M] ⇌ ROH ... RO—M—OH

ROH (top) / ROH (bottom) on arrows; R-O-M-O-H product

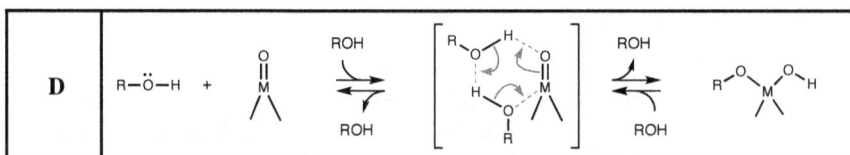

Overall, any reactions that can be written as shown below or contains similar structures may form by a series of proton transfer steps. These reactions take place in protic solvents and are therefore capable of following any of the mechanisms described above.

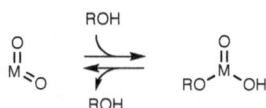

ROH (above arrow)

$$M(O)_2 \;\xrightleftharpoons[\text{ROH}]{}\; RO\text{-}M\text{-}OH$$

In order to simplify oxidations, the ester-oxidant formation step will not be shown. We will concentrate on the oxidation step. I suggest that you draw a plausible route for the formation of the ester-oxidant. Now that we have discussed the proton transfer reactions that take place in an oxidation reaction, let's move to the oxidation step, the second step of the reaction. In that step, a pair of electrons are transferred to the oxidant. For chromic acid oxidations, that step is rate limiting. The base could be water, but other bases are possible or it may be an intramolecular reaction.

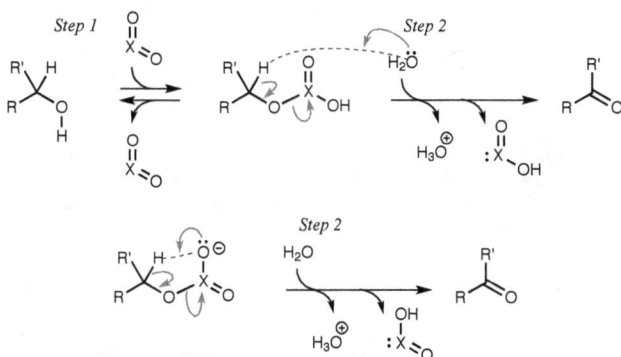

Step 1 ... *Step 2* ... H_2O ... $H_3O^{\oplus}$ ... $:X(O)_2OH$

Step 2 ... H_2O ... $H_3O^{\oplus}$ ... $:X(O)_2OH$

Because these oxidation reactions form esters with a metal oxide, I have written the structures of several common oxidants and their hydrated forms. The hydrated form is the product of water addition and any oxidation may be thought to involve either or mixtures of those forms (unless there is experimental evidence indicating otherwise).

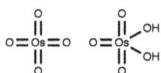

osmic acid

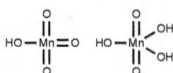

permanganic acid

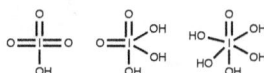

periodic acid

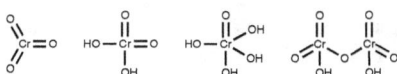

chromic acid

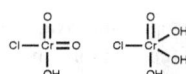

chlorochromic acid

Chromic Acid Oxidation

1. A chromic acid oxidation (Jones oxidation) of 3-methyl-2-butanol gives 3-methyl-2-butanone. As I indicated earlier, several proton transfer steps occur to form a chromate ester. In addition, the chromate ester is written in the hydrate form. This reaction is a compromise to make the steps appear logical, to highlight the actual oxidation step itself, and to not result in a lengthy series of proton transfer steps. I thought the hydrate form was in keeping with other oxidations in this chapter, though it may differ from your textbook. If you wish, you may write additional proton transfer steps to effect a loss of water to form an anhydro-ester (Cr=O plus H_2O to replace $Cr(OH)_2$).

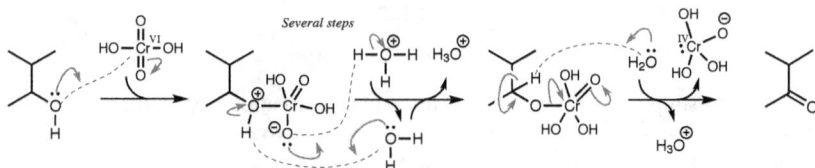

A variation that I think also looks plausible, is shown below.

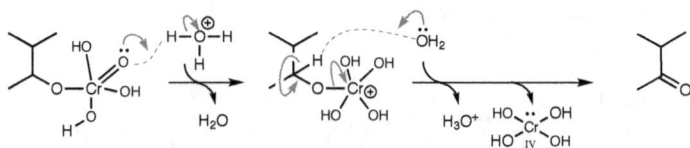

2. The chromic acid oxidation of isobutyl alcohol (2-methyl-1-propanol) gives isobutyric acid (2-methylpropanoic acid). Step 1 is the chromate ester formation and oxidation to an aldehyde.

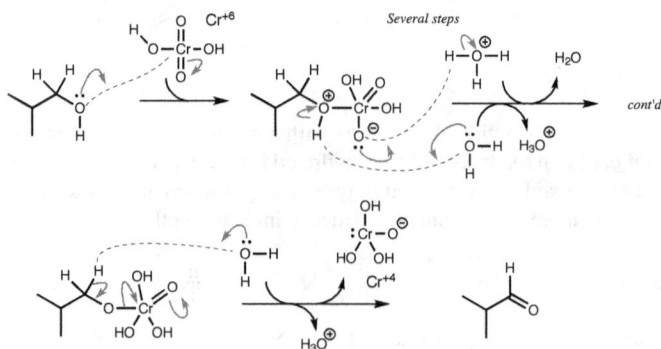

In Step 2, the oxidation step is exactly the same as in Step 1. However, the product of Step 1, an aldehyde, adds water to the carbonyl group (to form a hydrate). (You will need to complete the proton transfer steps.) The hydrate is another alcohol with a geminal hydrogen. The oxidation step is repeated to give isobutyric acid (2-methylpropanoic acid).

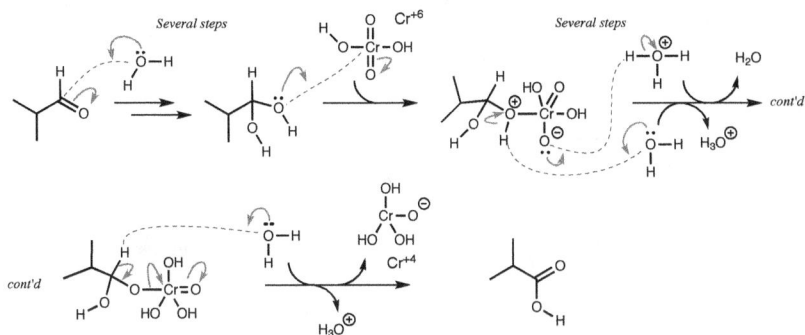

PCC, Tollens, Hypochlorite, *m*CPBA, and Sulfonium Based Oxidations

3. Pyridinium chlorochromate (PCC) oxidation of benzyl alcohol gives benzaldehyde. In order to isolate an aldehyde, the oxidation is carried out in the absence of water. Therefore, no hydrate of the aldehyde forms and the oxidation stops.

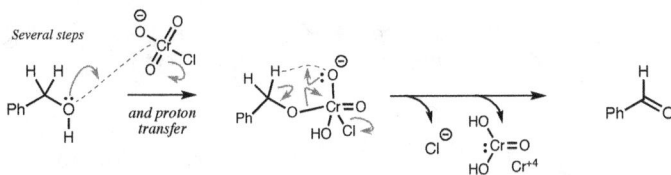

4. A Tollens oxidation of 2-methylpropanal (isobutyraldehyde) with silver oxide (or hydroxide) gives a carboxylic acid (after acidification).

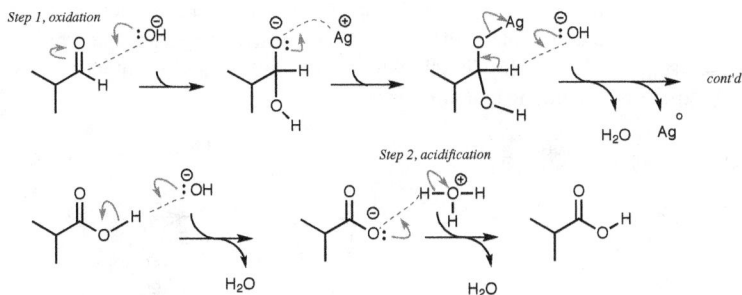

5. Cyclohexanol can be oxidized to cyclohexanone with sodium hypochlorite (NaOCl, bleach).

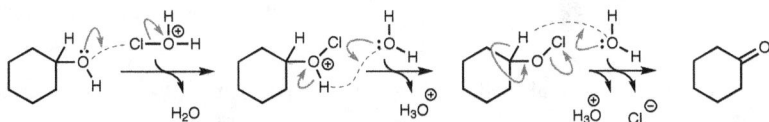

6. A peracid epoxidation of *trans*-2-butene with *m*-chloroperoxybenzoic acid (MCPBA) gives an epoxide, (2R,3R)-2,3-dimethyloxirane and its enantiomer. A challenge in writing rational mechanisms is sometimes it becomes imprudent to use discreet steps. While electron rich alkene react faster than electron poor ones, it is difficult to explain that with an epoxidation reaction. Thus, a mechanism with four pairs of electrons moving concurrently is suggested.

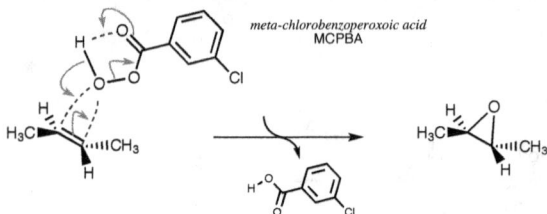

7. A Swern oxidation is a common laboratory oxidation method. The PCC and Swern oxidations can be carried out in anhydrous conditions and therefore result in aldehydes from primary alcohols. The Swern oxidation is a two step oxidation. In Step 1, a chlorosulfonium salt* is prepared by a reaction of oxalyl chloride and dimethyl sulfoxide (DMSO).

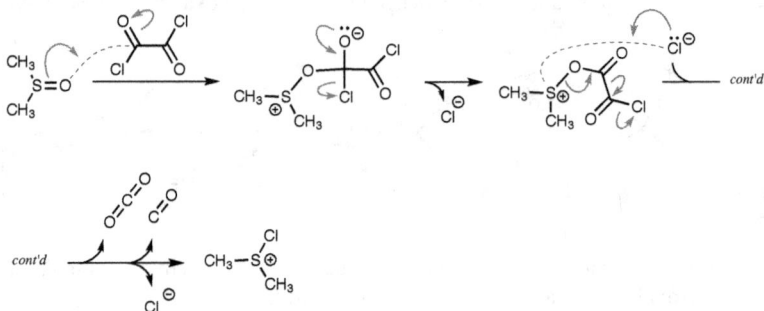

Step 2 of a Swern oxidation, oxidation of alcohol. After the reaction of the sulfonium salt with an alcohol, the oxidation steps converge. Some of the deprotonation/oxidation step may also occur directly with the triethylamine base. other alkyl groups have been used to avoid the malodorous dimethyl sulfide.

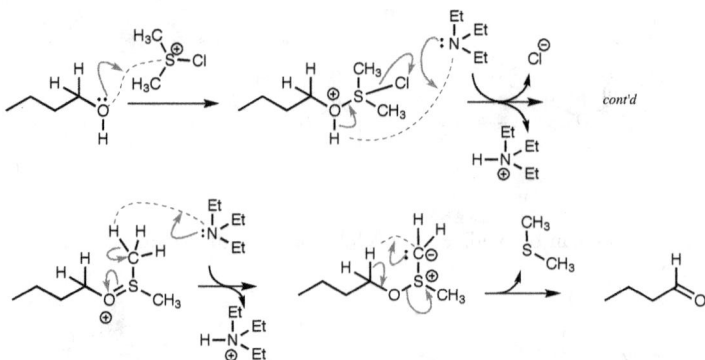

---

* This is one of several methods to form an equivalent sulfonium salt. Example 8 is another.

8. A Corey-Kim oxidation is another sulfonium salt based oxidation. The chlorosulfonium salt is prepared from N-chlorosuccinimide (NCS) and is used to oxidize cyclopentanol to cyclopentanone.

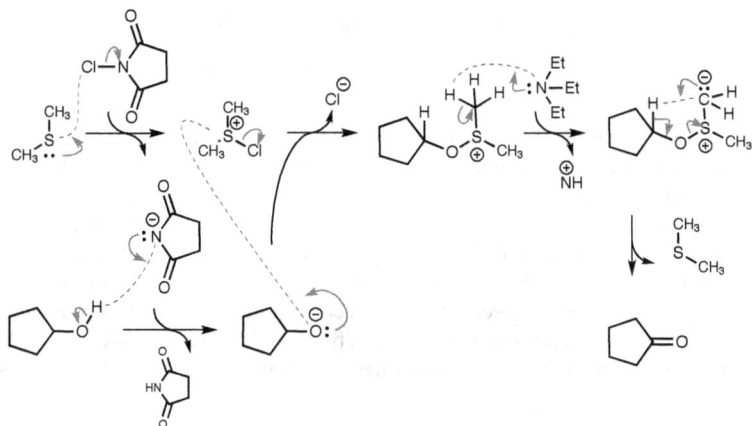

Ozone Oxidation

9. Step 1, ozone reaction with an alkene gives an ozonide. When I start the curved arrows, I start with the alkene and move them to give the most stable carbocation. The remaining bonds that form remove the formal charges. When I cleave the molozonide, I again start to give the most stable carbocation. The peroxyketone is its resonance structure.

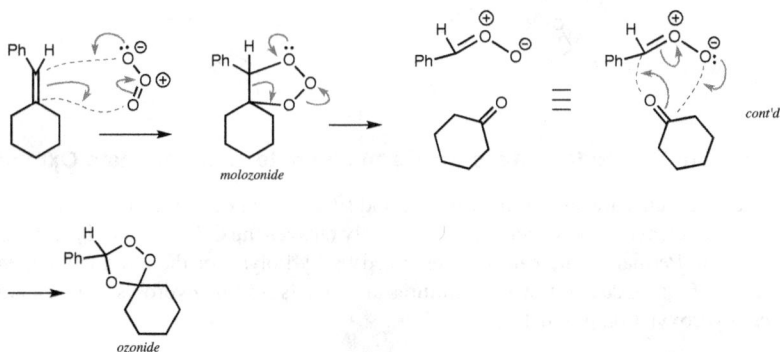

molozonide

cont'd

ozonide

Step 2, reduction of ozonide to two carbonyl compounds. You can use other reductants than dimethyl sulfide. I used it here to avoid an extra step to go from a hydrate to a ketone. An ozonide is actually a peroxy diacetal. For reductive Step 2, two goals are accomplished, reduction of the O-O bond and conversion of the acetals or ketals into two carbonyl groups. If you use zinc, you will also reduce the O-O bond and then hydrolyze the mixed hemi-ketal/acetal. The acetal hydrolysis (see Chapter 8) will require more steps than dimethyl sulfide.

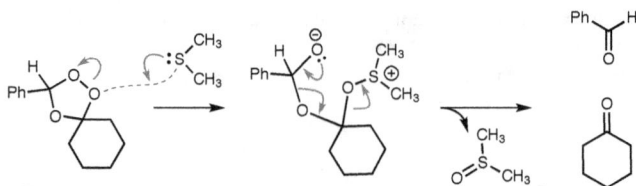

Alternate Step 2, oxidative destruction of ozonide. If the double bond has a hydrogen attached, the C-H bond electrons can be used to break the O-O bond as in other oxidation reactions (or see Chapter 6.3). In alternate Step 2, the O-O bond is used to convert an acetal to a carboxylic acid. If a dicarboxylic acid could be formed, then an additional mole of hydrogen peroxide would be required. Since the oxidation of an ozonide is related to the Baeyer-Villiger oxidation (Chapter 6.2 and 6.3), Baeyer-Villiger products are frequent by-products. It is sometimes advantageous to separate the second oxidation step to avoid such by-products, that is, first do a reductive work-up, isolate the aldehydes and ketones, and then oxidize the aldehyde in a separate step.

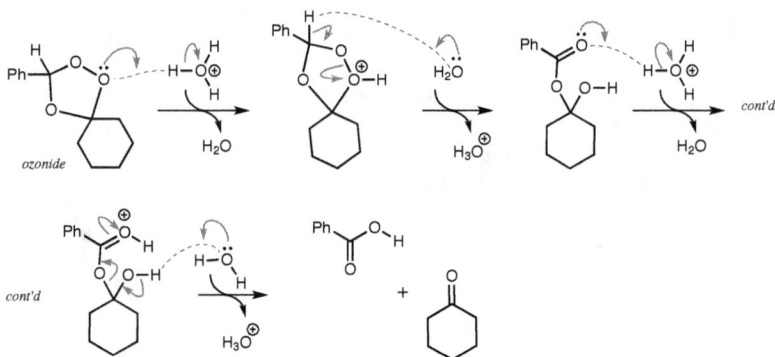

ozonide

cont'd

cont'd

Osmium Tetroxide, Potassium Permanganate, and Periodate Oxidations

These reagents are similar in structure and function. They differ in that osmium tetroxide does not cleave the C-C bond, periodate only cleaves the C-C bond, and permanganate can do both. Permanganate can be intercepted by hydrolysis of the intermediate cyclic ester. Excess hydroxide and low temperatures are signals for that hydrolysis and isolation of the cis-hydroxylation product.

10. Osmium tetroxide oxidation of cyclohexene gives cis-1,2-cyclohexanediol. The first arrow is the oxidation and the second arrow is several steps for a hydrolysis of the osmate ester. The anhydride form (OsO_3) was written to remain in the same form as the regeneration step, but it it could also have been written as the hydrate, H_2OsO_4.

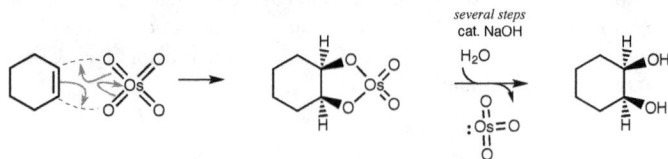

Regeneration of oxidant with *tert*-butylperoxide. This permits the conversion of an alkene to a *cis*-diol with a catalytic amount of osmium tetroxide.

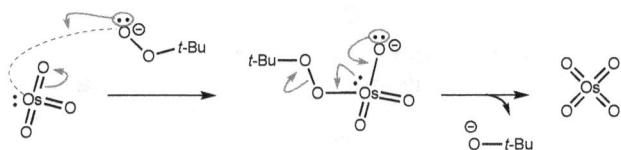

11. Potassium permanganate can oxidize methylcyclohexene to 1-methylcyclohexane-1,2-diol at low temperature or to a keto-acid at higher temperatures. This oxidation is similar to the osmium oxidation and forms a manganate ester. This ester can be hydrolyzed at low temperature to isolate a *cis*-diol.

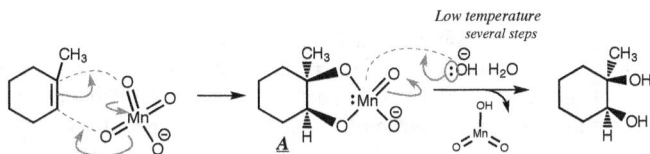

At higher temperatures, the manganate ester (**A**) will decompose with cleavage of the C-C bond. If the double bond contains a hydrogen atom, then the oxidation will continue to a carboxylic acid. In the example, an intermediate aldehyde is formed. This can add water and form another manganate ester from the hydrate-OH. This ester will decompose to a carboxylic acid. The oxidation forms hydroxide which forms the salt of the carboxylic acid, therefore, a step in a permanganate oxidation, if heated, must be a neutralization step. If it is the final step, then neutral impurities can be easily removed.

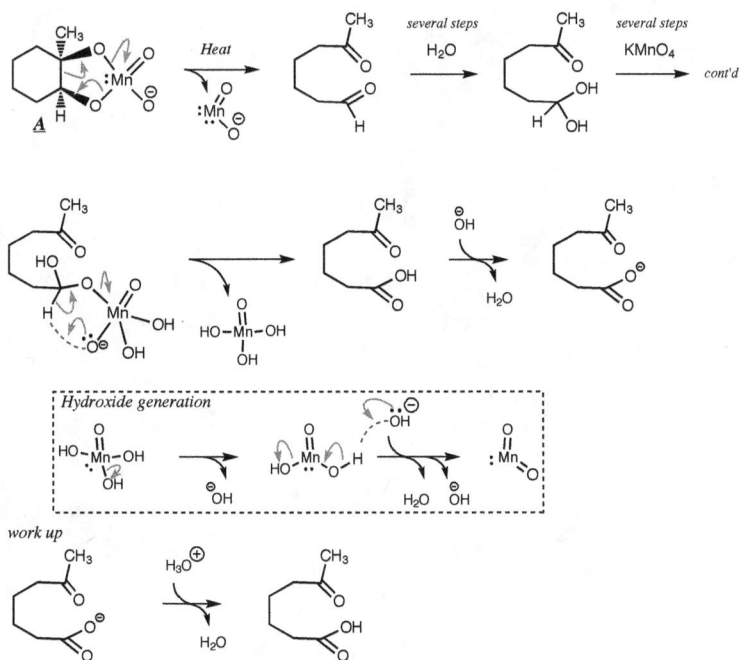

The final disposition of manganese (and other metals) is more complex than shown here. Potassium permanganate can interact with manganese intermediates as well as auto-decompose when heated.

Permanganate, when heated can be used for side chain oxidations of aromatic compounds. . The oxidation of ethylbenzene to benzoic acid first forms benzylic radical. This oxidation may include the two electron oxidations, addition of water, and proton transfers consistent with the oxidation steps shown in this chapter.

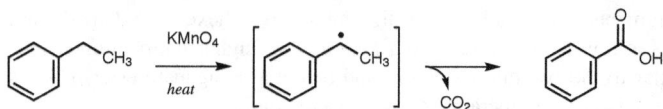

12. Periodate completes these similar oxidations as it will cleavage a 1,2-diol to give a dicarbonyl compound. The periodate ester steps have not been shown, but they are the reverse of the osmate or permanganate hydrolysis steps. If the alcohols can be contained in a five-membered iodate ring, then the oxidation will occur.

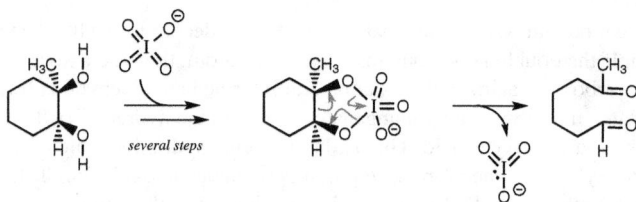

13 — Organometallic Reactions

The (transition) organometallic reactions in this chapter are of increasing importance. Organometallic chemistry often involves reactions in which the mechanism is not understood, incompletely understood, or complex. In some reactions, I show how it *might* take place. The reactions may recount the steps more than explain a reaction. Significantly absent is the concept of 18 valence electrons in a description of organopalladium chemistry.

Acyclic Heck Reaction

1.   The Heck reaction is a very useful reaction because it only requires a catalytic amount of palladium, the reaction conditions are very mild, and a wide variety of functional groups are compatible with the reaction. We will study two similar examples, reactions of acyclic and cyclic alkenes. Before the coupling reaction takes place, zero valent palladium metal, the active catalyst, must be formed in a sub-reaction. Palladium acetate reacts with an alkene (much like mercuric acetate, but in an *anti*-Markovnikov orientation). The intermediate is unstable and loses hydridopalladium acetate, which in the presence of base forms ammonium acetate and zero valent palladium (Pd^0). Zero valent palladium is commercially available and when used, this sub-reaction is skipped.

Step 1, reduction of palladium (II) to zero valent palladium with propene. The steps in the Heck reaction seem fairly well understood. This step may be skipped if a zero valent palladium catalyst is used, however if you wish to learn the chemistry of the reaction, it compliments the coupling steps and is useful for this reason.

Also important to the reaction are the ligands that donate electrons to palladium. Palladium chemistry is consistent with formation of an 18-electron intermediate. Therefore the ligands, although not included in the mechanism, are vital to the success of the reaction. The electrons of an alkene presumably replace the electrons of one of the ligands that coordinate with the palladium, see box.

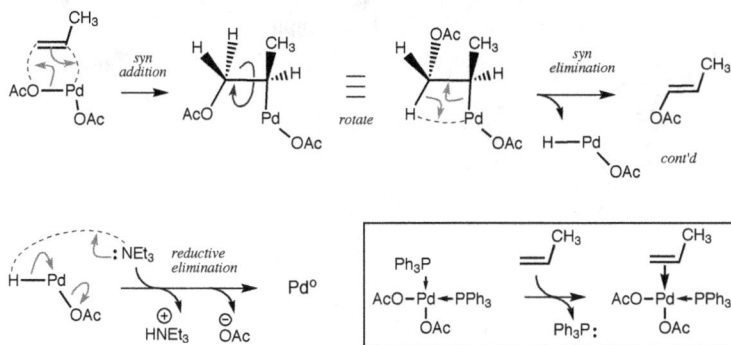

In Step 2, the coupling sub-reaction, palladium is inserted into a carbon-bromine bond. This reaction is similar to the insertion of magnesium into a carbon-halogen bond to form a Grignard reagent. This reaction converts palladium from an electron donor to an electron acceptor. The rest of the process is similar to the palladium reduction sub-reaction above. The organopalladium compound inserts into an alkene double bond. The alkene pi-electrons probably complex and are donated to the palladium before the *syn*-addition takes place. In order for an elimination to occur, the bonds must rotate. A *syn*-elimination gives the coupled product and hydridopalladium bromide. The hydridopalladium bromide can lose hydrogen bromide and regenerate palladium zero.

Step 2, the catalytic cycle (oxidative-addition, *syn*-addition, *syn*-elimination, and reductive-elimination) with 2-bromopropene and propene.

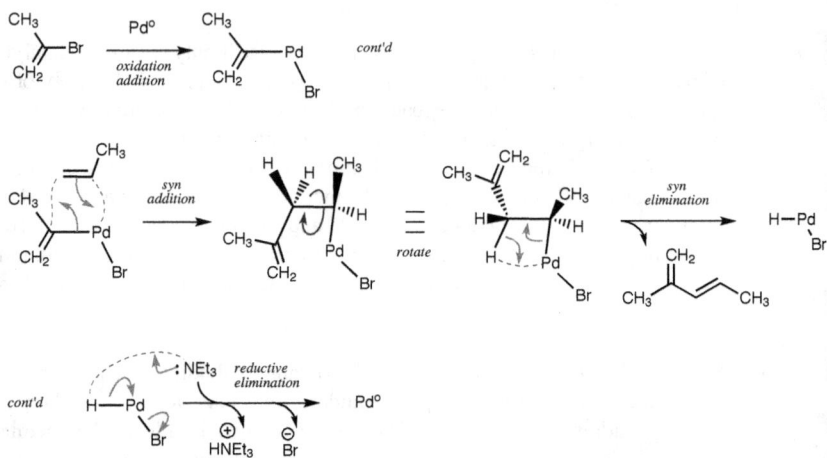

Cyclic Heck Reaction

2.   The chemical steps are the same as in Example 1, but the stereochemistry alters the products that form. Step 1, reduction of palladium (II) to zero valent palladium with cyclopentene. The *syn*-elimination reaction must occur on the opposite side as the acetate. The acetate is now *cis* to the palladium and the ring prevents a bond rotation. In Example 1, a vinyl ester formed. In Example 2, a cyclic allylic ester forms.

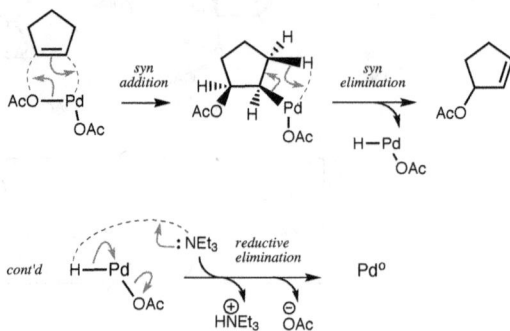

Step 2, the catalytic cycle (oxidative-addition, *syn*-addition, *syn*-elimination, and reductive-elimination) with iodobenzene and cyclopentene.

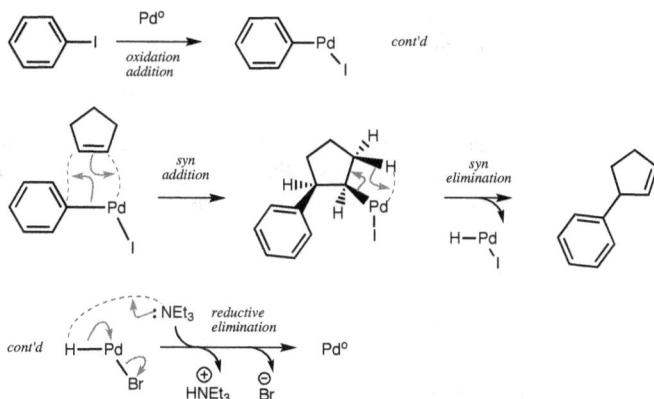

Catalytic Reduction of an Alkene

3. This mechanism is hypothetical. I extrapolated from the Heck reaction to similar mechanisms and other metals. This mechanism can broadly explain formation of the products and by-products of catalytic reduction reactions. It also explains how *trans*-fatty acids can be produced. Catalytic hydrogenation of *cis*-3-hexene gives hexane. If there isn't enough hydrogen to complex with the palladium, then the *cis*-addition can reverse. If you started with a *cis*-alkene, you could end up with the *trans*. This is how *trans*-fatty acids are formed in vegetable oils.

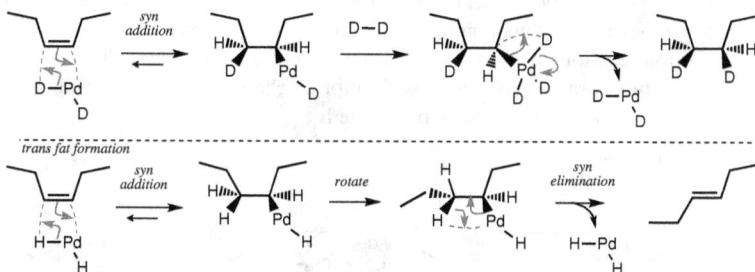

Gilman Reagent

4. This mechanism is also hypothetical. The mechanism of the Gilman coupling is not known. I've written it to broadly agree with how these reagents *might* react. The groups couple without losing the stereochemistry. While I don't know how that might occur, I anticipated that just as zero-valent palladium could insert into a carbon-halogen bond, so too might copper insert into a carbon-bromine bond. If it did, then the intermediate on the right would occur. The decomposition of this intermediate parallels that of the palladium (Heck) and mercury (reduction step of oxymercuration) reactions.

In a Gilman reagent, replacing an iodide of copper iodide with a methyl group should make copper a better electron donor. If second methyl group were also added to the copper, this copper should be an even better electron donor.

Formation of Gilman reagent, lithium dimethylcuprate.

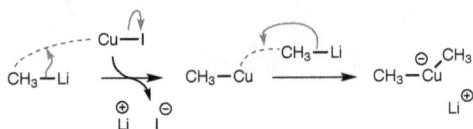

Coupling of Gilman reagent, lithium dimethylcuprate with (Z)-1-bromopentene to give *cis*-2-hexene. Each of these reagents, Gilman, magnesium, lithium, and palladium, presumably donate electrons to break the carbon-halogen bond and gives a net insertion.

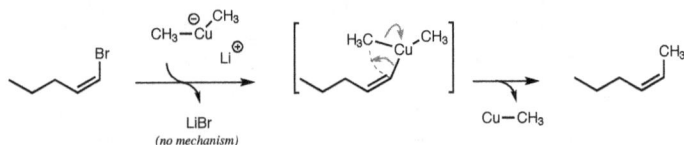

LiBr
(no mechanism)

If you look back to the de-mercuration step of the oxymercuration reaction (Chapter 5.16) this is reported in the literature to be a radical reaction. The decomposition of the alkylcuprate has also been suggested to be a radical reaction. However, I like to write this as a two-electron process as it is similar to other reactions in this book, e.g., oxidation of boranes, Baeyer-Villiger oxidation, loss of PdH$_2$, and others. I like the two-electron mechanism as this nicely parallels these other reactions. By using a parallel, they are more plausible, easier to understand, remember, and use (which is an objective of this book).

5. A 1,4-conjugate addition of lithium dimethylcuprate to 2-cyclohexenone gives 3-methylcyclohexanone. This conjugate addition could be consistent with other conjugate addition reactions if the Gilman reagent is not a strong base. The decomposition of the organocuprate is consistent with prior mechanisms. The intermediate enolate can be alkylated as well as simply protonated.

Step 1, addition   *Step 2, Work up*

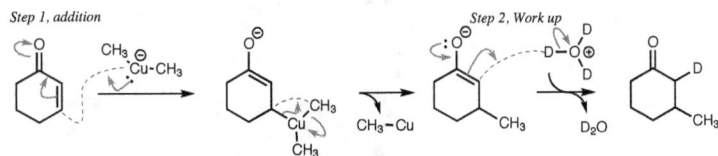

6. Reaction of lithium dimethylcuprate to benzoyl chloride to give acetophenone. The formation of a ketone product is also consistent with the Gilman reagent not being strongly basic, therefore it does not add to the ketone carbonyl group as a Grignard reagent would.

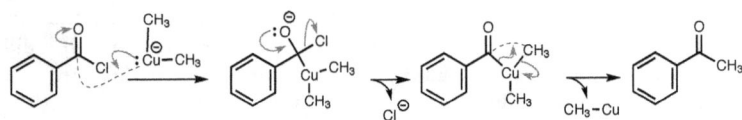

14 — Aromatic Substitution Reactions

Electrophilic Aromatic Substitution of Benzene

1. Friedel-Crafts acylation of benzene.

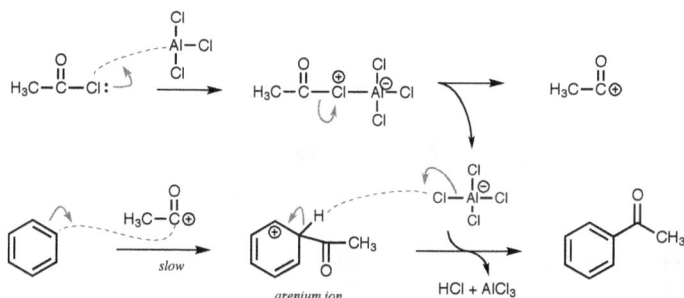

2. Friedel Crafts alkylation of benzene. This is only one of many variants of Friedel-Crafts alkylation reactions. If a secondary or tertiary alkyl group is to be attached, then an alcohol or an alkene can also be used to generate a carbocation intermediate.

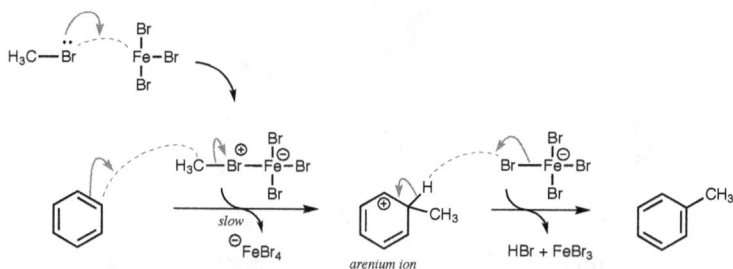

Friedel-Crafts alkylation reactions with primary halides occur with extensive rearrangement. In order to prepare *n*-butylbenzene via an electrophilic aromatic substitution, the common route is to first do a Friedel-Crafts acylation reaction and reduce the product with a Clemmenson or a Wolff-Kishner reduction. The major product of a direct Friedel-Crafts alkylation is a rearranged product. Also, since the product is more reactive than the starting material, mixtures may result.

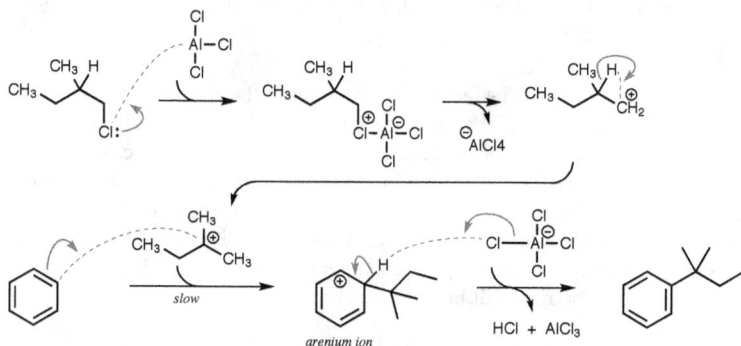

3. Ferric bromide bromination of benzene. A Lewis acid is normally necessary to increase the reactivity of bromine. However, with very reactive aromatic rings, such as phenol, it can be brominated without a catalyst.

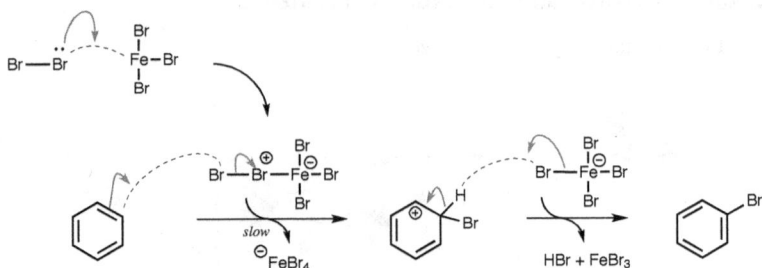

4. Nitration of benzene with nitric and sulfuric acids.

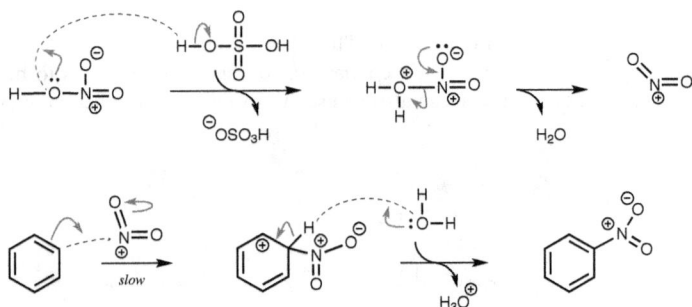

Electrophilic Substitution of Substituted Aromatic Compounds

Substituents on a benzene ring can direct the electrophilic aromatic substitution reaction. There are two general methods these problems are solved. In one method, the resonance structures of the arenium ion intermediates are drawn for *ortho*, *meta*, and *para* attack. Those resonance structures are analyzed for structures that may be stabilizing or destabilizing. However, I am dissatisfied with this method in that the resonance structures of the arenium ion intermediates infers the electrophilic attack is reversible and that the extra stabilization or destabilization shifts an equilibrium to the favored isomer. It also does not indicate which reaction should be more facile than others.

Another method is to draw the resonance structures of the substituted aromatic compounds. The resonance structures of the reactants can be used to predict whether the atoms of a benzene ring will be **more** or **less** electron rich in an electrophilic aromatic substitution reactions. I like this method of examining the resonance structures of the reactants to predict the regiochemistry of the reaction. I believe the reaction profile is more consistent with the slow step being addition to an aromatic ring (and disrupting aromaticity). I am also more satisfied by how the resonance structures predict reactivity because those which add electron density to the ring will be more reactive and those which remove electron density are less reactive.

However, you will find that halogens do not follow that pattern. They contribute electrons by resonance and are thus *ortho-para* directors, but are less reactive. However, I rationalize that although halogens have non-bonded electrons to donate, but because of their electron withdrawing properties, are reluctant to do so.

Let us look at the general principle. First, draw the resonance structures for the substituted benzenes, shown below. Examine the resonance structures and answer; *"Will the compound react faster or more slowly than benzene in an electrophilic substitution reaction (such as bromination or nitration)? Where on the molecule will the highest electron density be found? At the ortho-para or meta position?"*

General model of electron donating groups (usually, X = N, O, halogen) are *ortho-para* directors. The *ortho* and *para* positions are electron rich positions in the resonance structures.

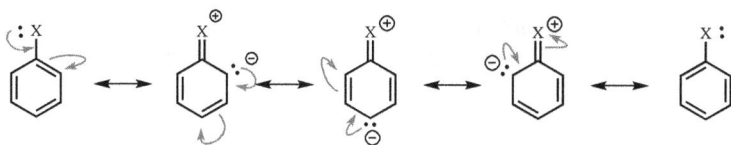

General model of electron withdrawing groups (usually Y = C, N, P, or S; Z = N, O, or S) are *meta* directors. The *ortho* and *para* positions are electron deficient. The *meta* position is not electron rich, just not deficient.

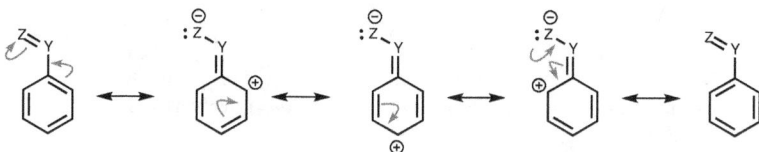

Examples for predicting site of electrophilic aromatic substitution

• Anisole, *ortho-para* director

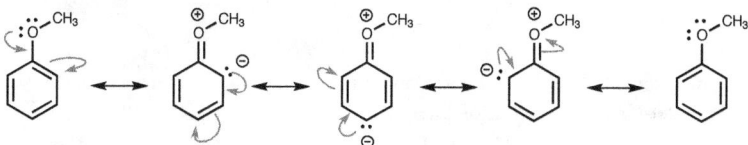

• Acetophenone, *meta* director

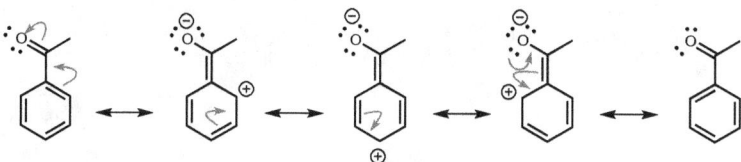

- Toluene, weak *ortho-para* director

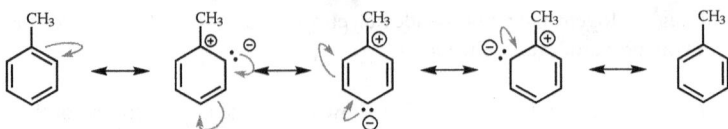

- Benzonitrile, *meta* director

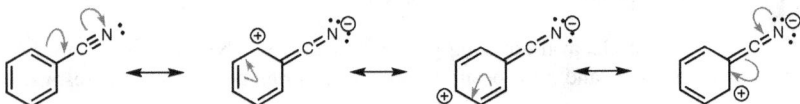

5. The acetyl group is an electron withdrawing group and this will react more slowly than benzene. It is also a *meta* director. Bromination of acetophenone gives *m*-bromoaceto-phenone.

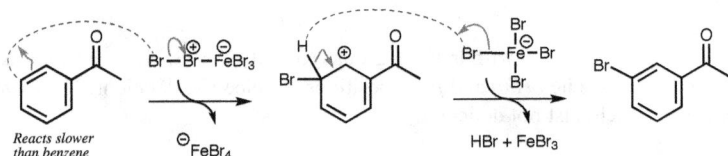

Reacts slower than benzene

$^{\ominus}FeBr_4$

$HBr + FeBr_3$

6. A carbomethoxy group is electron withdrawing, a *meta* director, and slower reacting. Nitration of methyl benzoate gives methyl *meta*-nitrobenzoate.

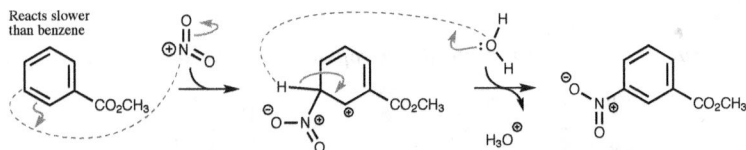

Reacts slower than benzene

$H_3O^{\oplus}$

7. An ester of phenol is electron donating, an *ortho-para* director and faster reacting. Friedel-Crafts acylation of phenyl acetate with acetyl chloride and aluminum chloride. If a substituent is an *ortho-para* director, it is difficult to predict whether the major isomer will be *ortho* or *para*.

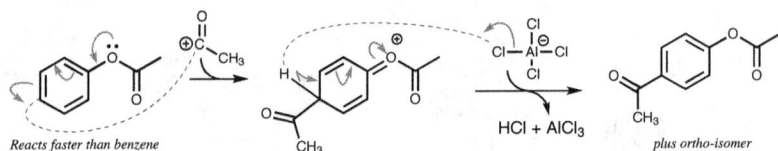

Reacts faster than benzene

$HCl + AlCl_3$

plus ortho-isomer

8. Triflic acid catalyzed acetylation of toluene gives *o*- and *p*-methylacetophenone.

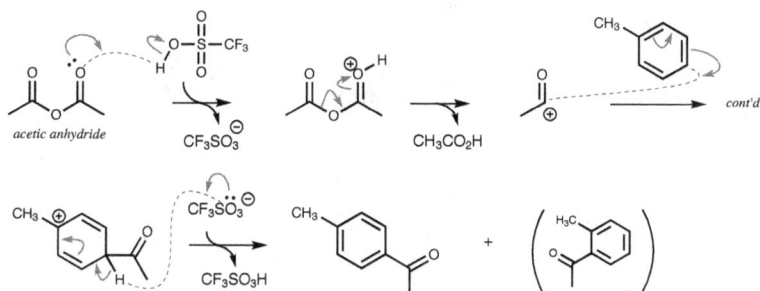

Aromatic Substitution on Multi-substituted Aromatic Rings

If a multi-substituted benzene reacts, the **activating substituents have the greatest effect on determining the regiochemistry of the products**. In the methoxybenzoic acid shown below, the methoxy group has the greatest effect on the substitution pattern whether the carboxy group is *ortho*, *meta*, or *para* to it.

The resonance structures of *p*-methoxybenzoic acid. Substitution is preferred *ortho-para* to the methoxy group and *meta* to the carboxy group.

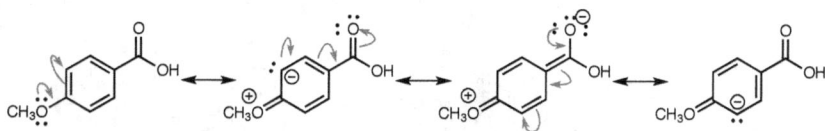

The resonance structures of *m*-nitroanisole. Substitution is preferred *ortho-para* to the methoxy group and *meta* to the nitro group.

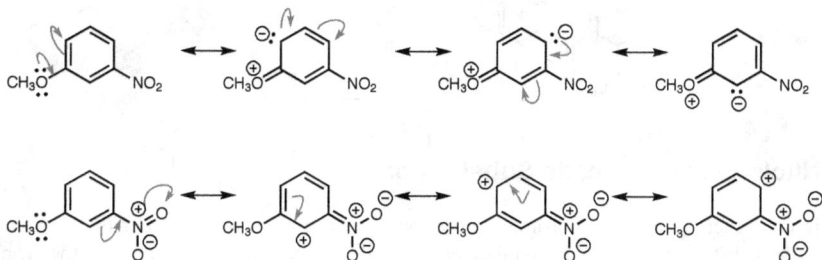

In *m*-nitroanisole, the CH₃O-group is activating and an *ortho-para* director but because of the nitro group, it is of reduced reactivity and regioselectivity. The upper series of resonance structures show where electron density will be increased. Substitution will be best at those positions. The lower series of resonance structures shows the effect of the nitro group. It will not control the regiochemistry, but it will decrease the electron density (and thus reactivity) created by the methoxy group.

9. Aluminum chloride catalyzed chlorination of methyl 3-methoxybenzoate (methyl *m*-anisate) gives methyl 2-chloro-5-methoxybenzoate.

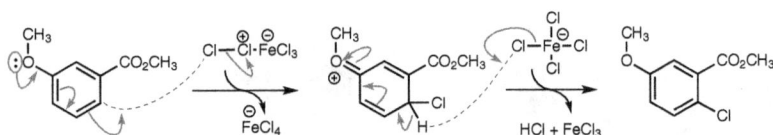

10. A Friedel-Crafts alkylation of 4-nitro-*N-p*-tolylbenzamide with two equivalents of chloromethane. In the reaction of the second equivalent of chloromethane and aluminum chloride, no reaction occurs in the *p*-nitrobenzamide ring. Since Friedel-Crafts reactions are the least reactive of electrophilic reagents, no reaction occurs in the ring with the two electron withdrawing groups.

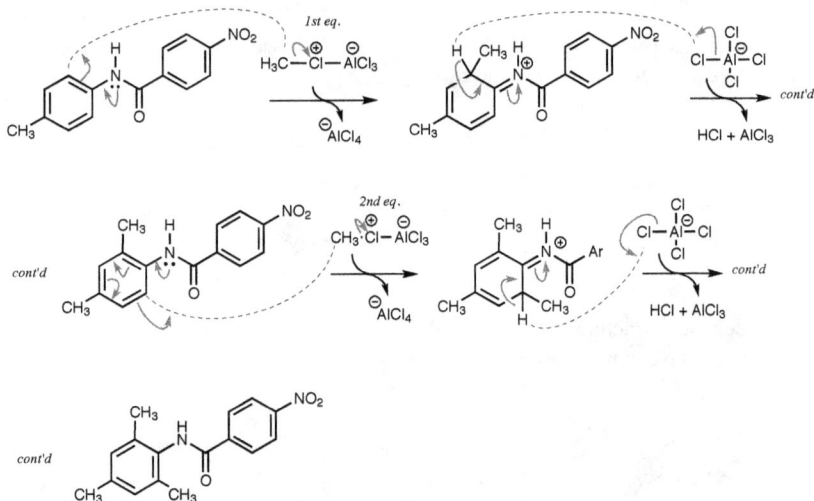

Nucleophilic Aromatic Substitution

If you react a substituted aromatic compound with a nucleophile rather than an electrophile, a different reaction mechanism ensues. The nucleophile can only add to the aromatic ring provided the aromatic ring contains enough electron withdrawing groups. The electron withdrawing groups must be able to absorb the negative charge of the nucleophile. The reaction is an addition-elimination reaction in a manner similar to an acid chloride. The most effective electron-withdrawing group is a nitro group, but other groups will work at slower rates.

11. Nucleophilic aromatic substitution of 1-fluoro-4-nitrobenzene with ammonia gives 4-nitroaniline. The nitro group has absorbed the negative charge in the second structure, a Meisenheimer complex.

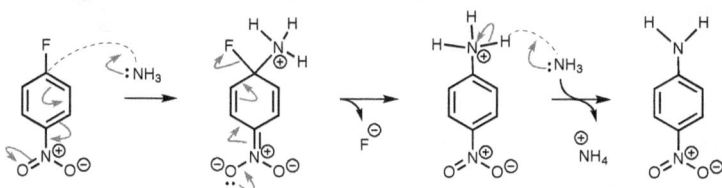

12. Nucleophilic aromatic substitution of 1,2-difluoro-4-nitrobenzene with sodium methoxide gives 2-fluoro-1-methoxy-4-nitrobenzene. Only one fluorine can be substituted. The fluorine in which the electrons of a nucleophile can be absorbed by the nitro group.

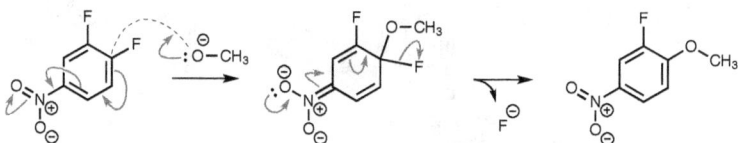

Benzyne Reaction

an addition-elimination reaction can take place between a nucleophile a halobenzene if there is a strong electron withdrawing group present to accept the electrons. However, if no electron withdrawing group is present and the nucleophile is a sufficiently strong base, a deprotonation reaction may take place that can result in an elimination reaction to form a benzyne. The benzyne, because the triple bond is in a six-membered ring, is very reactive. Thus, a nucleophile can add to the triple bond at either end and the regiochemistry of the original halobenzene becomes lost.

13. Reaction of 4-bromoanisole with sodium amide (sodamide, NaNH$_2$) gives 3- and 4-methoxyaniline via a benzyne intermediate.

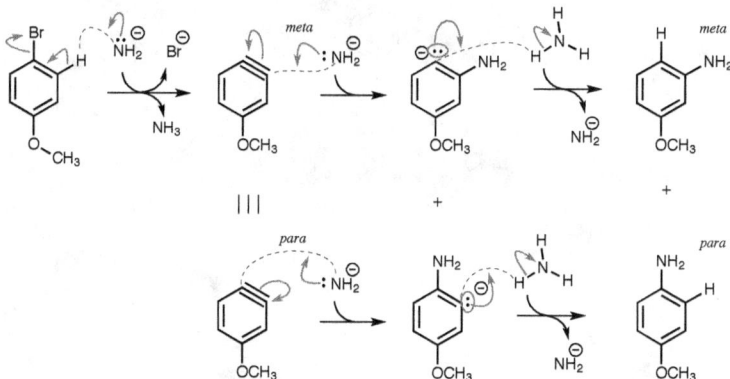

Diazonium Chemistry

14. The diazotization reaction is not as complicated as it may look. It has six proton transfer reactions (acid-base chemistry), two dehydration reactions, and a condensation
of NO⁺ with aniline. One of the steps is like an acid catalyzed enolization except the
carbon atoms are replaced by two nitrogen atoms. However, the most frequent error
that I have noted has been due to an incorrect Lewis structure for sodium nitrite.

Formation of a diazonium salt from aniline.

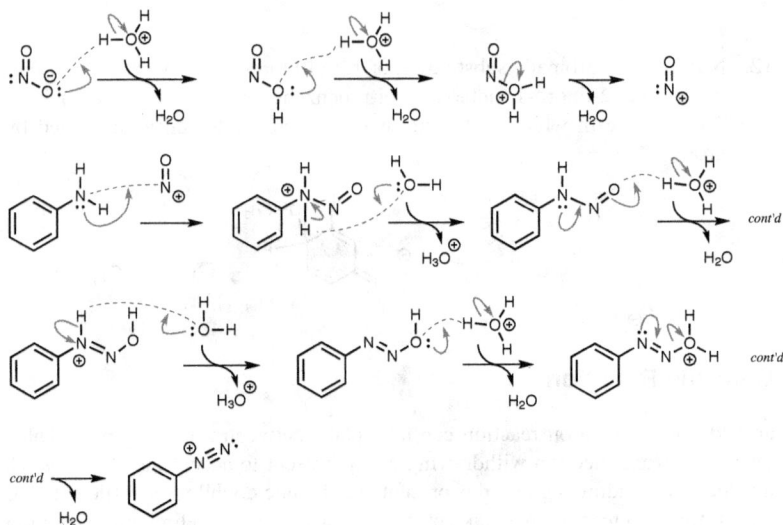

Reaction of diazonium salts. This is not a mechanism. It is just a list of some of the
products that can form.

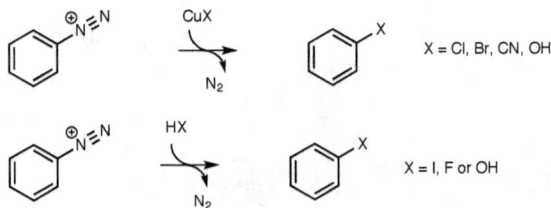

X = Cl, Br, CN, OH

X = I, F or OH

15 — Carbene and Nitrene Reactions

Carbene Reactions

This is a simplified model of carbene chemistry. There are two types of carbenes, singlets and triplets, and they have different rates of reactivity and stereochemical outcomes. In general, singlet carbenes are more reactive and experience less loss of stereochemistry than an equivalent triplet carbene.

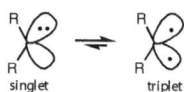

singlet   triplet

The two forms can interconvert. The singlet is closer to a carbocation in its characteristics. It has a vacant orbital which it fills upon reaction. A triplet has two half-filled orbitals. When it reacts with an alkene, it reacts as a diradical and can lose its stereochemistry. Often, the triplet is the more stable form.

1. The Simmons-Smith reagent has great variety in it reactions. In this example, the Simmons-Smith carbene adds to cyclohexene to give a bicyclo[4.1.0]heptane. Your textbook may use a concerted addition of the carbene or carbenoid group to an alkene. I prefer this alternate mechanism as it does not violate the symmetry rules for a 2+2 reaction, and it follows the pattern of bromine, mercury, or borane additions to an alkene in Chapter 5. I also like it because it is similar to Example 2 and can explain the reaction of a Simmons-Smith reagent as a nucleophile in other reactions (not shown),

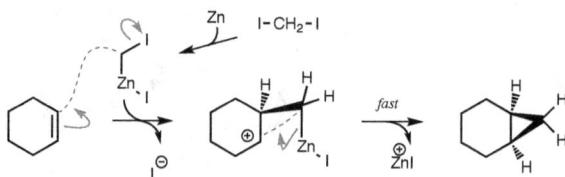

2. A dihalocarbene addition to (E)-1-phenylpropene to give a dibromocyclopropane.

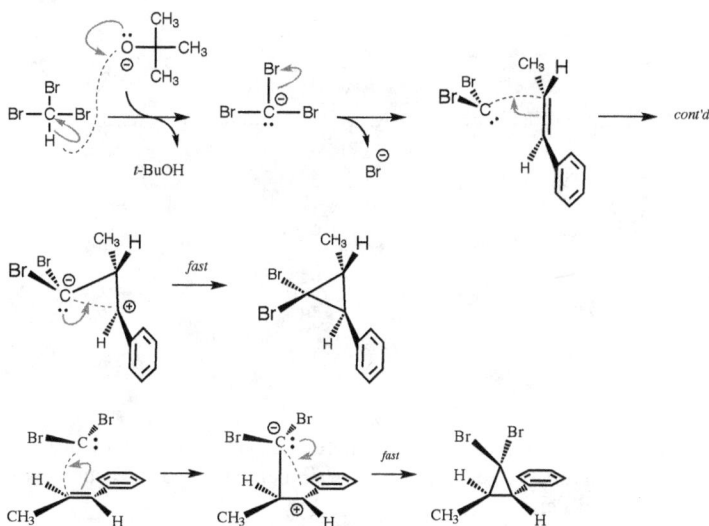

3.   A dihalocarbene addition to cyclohexene to give 7,7-dichlorobicyclo[4.1.0]heptane.

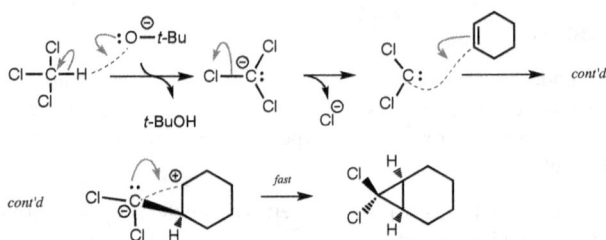

Curtius Rearrangement

4.   Step 1, Curtius rearrangement, reaction of acid chloride with azide and rearrangement via a nitrene intermediate.

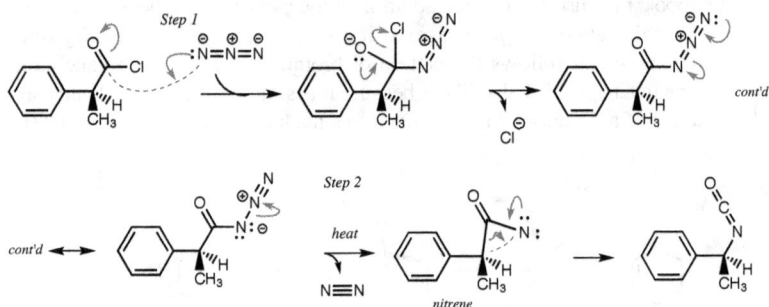

Step 2, hydrolysis of isocyanate to an amine.

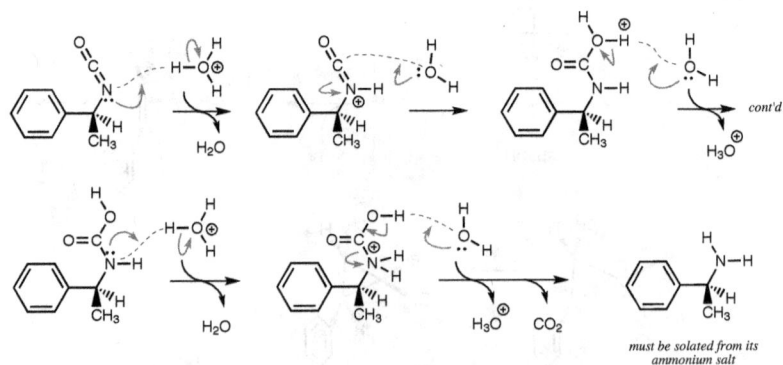

Alternate Step 2, hydrolysis of isocyanate to a carbamate. A carbamate is a nitrogen analog of a carbonate ester. The steps are similar to the above reaction with an alcohol replacing the water.

methyl hydrogen carbonate

methyl phenylcarbamate

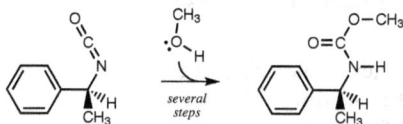

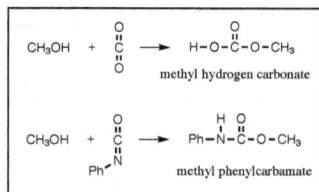

Hoffmann Rearrangement

5. Step 1, the reaction of the amide with sodium hydroxide and bromine in a Hoffmann rearrangement of a primary amide to give the corresponding amine.

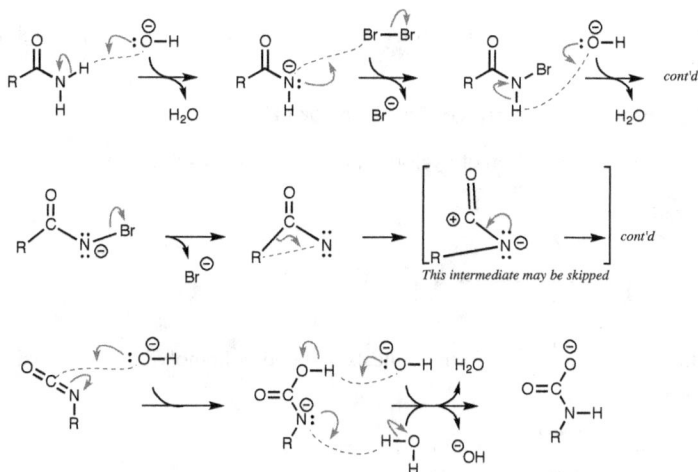

This intermediate may be skipped

Step 2, Acidification and decarboxylation to give the amine.

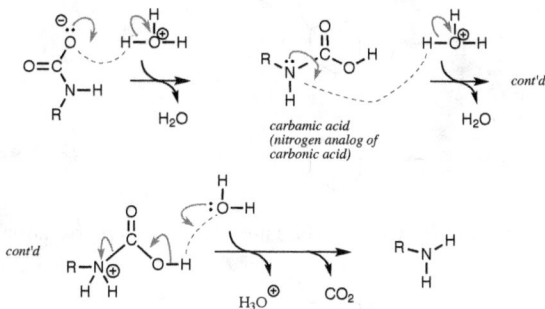

carbamic acid
(nitrogen analog of carbonic acid)

16 — Radical Reactions

I placed radical reactions in the last chapter. While radical reactions should not be thought of as unusual, you should also note that the driving force for their reactivity is to form paired electrons. While the remainder of the book has been about reactions of paired electrons, that is electron pairs being attracted to nuclei, I wanted radical reactions to be in a chapter of their own. Electrons being attracted to electrons to form an electron pair is a different kind of attraction and reaction. Also, single electron movements are written differently from electron pair movements, they use a single barbed arrow. You should consult your textbook for information on rates and selectivity of radical reactions.

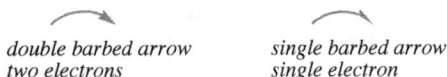

double barbed arrow
two electrons

single barbed arrow
single electron

Chlorination and bromination reactions are mechanistically the same, though they have different selectivity.

Free Radical Bromination (or Chlorination) Reaction

1. Free radical bromination of cyclohexane gives bromocyclohexane.

    Overall reaction

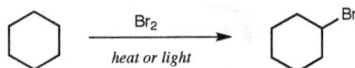

    Br_2, heat or light

    Initiation, the first step, forms radicals. A bond is homolytically cleaved by heat or light.

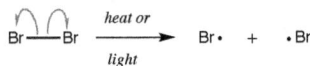

    $Br{-}Br$ $\xrightarrow{heat\ or\ light}$ $Br\cdot$ + $\cdot Br$

    Propagation, the transfer of the radical to form new radicals and the product.

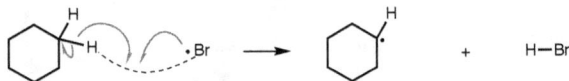

    + $H{-}Br$

    + $\cdot Br$

    Termination, forms a bond from two radicals. This ends a chain propagation reaction.

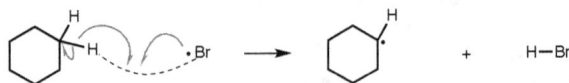

    + $H{-}Br$

    + $\cdot Br$

Allylic Bromination with NBS

2. Free radical bromination of cyclohexene with *N*-bromosuccinimide, an allylic bromination.

Overall reaction

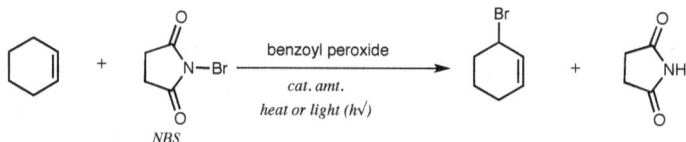

Initiation was understandable when bromine was present and bromine radicals are part of the propagation reactions. In many other reactions, another chemical is used to form radicals which become transferred to the bromine radicals.

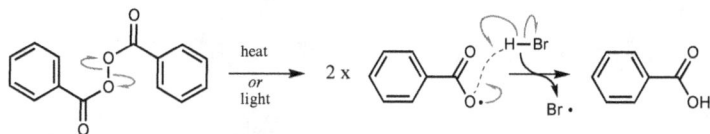

Propagation in an NBS reaction requires another step to from low levels of bromine. It may also be considered as to absorb HBr liberated from the propagation reaction.

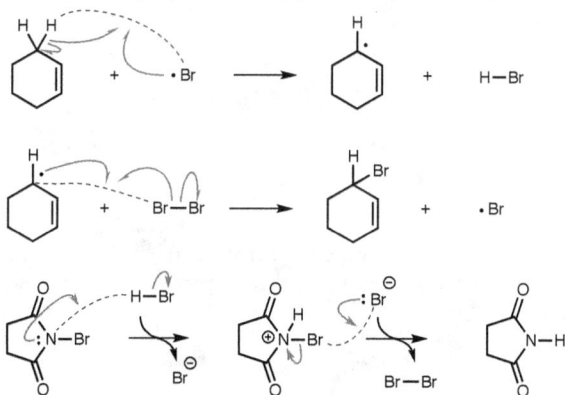

Termination

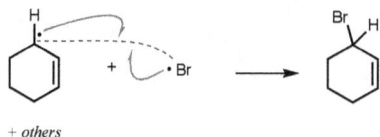

+ *others*

Several catalysts are capable of initiating radical reactions. Benzoyl peroxide and azobisisobutyronitrile (AIBN) are commonly used. While the reaction generates an oxygen radical to start the reaction, the unpaired electron becomes transferred to a bromine radical. The following is a possible path to generate a bromine radical from an oxygen radical.

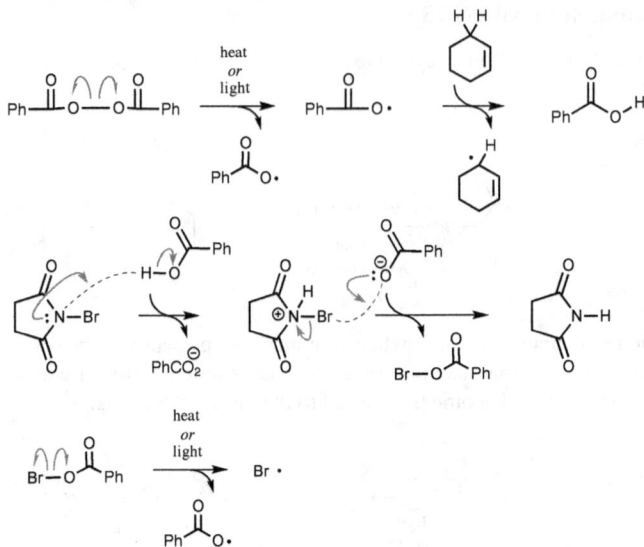

Radical Addition of Hydrogen Bromide

3. Free radical addition of HBr/H$_2$O$_2$ to propene gives 1-bromopropane via an *anti*-Markovnikov addition of hydrogen bromide.

    Overall reaction

    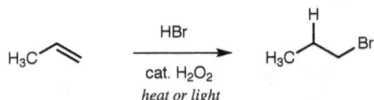

    Initiation does not require a full equivalent of peroxide. It is only present to generate bromine radicals. Here, I have written a sequence to generate them from the hydroxy radical.

    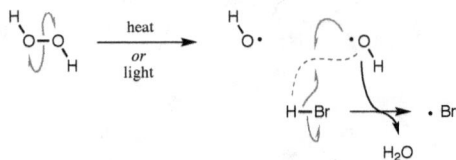

    Propagation

    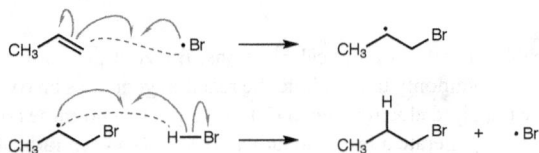

Termination

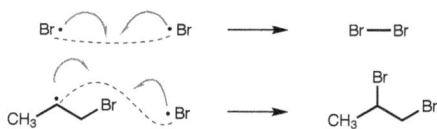

others

Why does NBS lead to an allylic bromination while a radical addition of HBr leads to addition? Both are reactions of a bromine radical with an alkene. If addition of a bromine radical is the expected product, how can one get allylic bromination (eq 2) to occur if the addition of a radical to a terminal position (eq 1) is a faster reaction? If the addition of the bromine radical to a terminal alkene is a fast but reversible reaction, then its success would be aided by quenching the intermediate radical with HBr. If the concentration of HBr is low or absent, then the reaction may reverse. Then a competing abstraction of the allylic hydrogen in a bromination reaction might occur. The use of *N*-bromosuccinimide is effective as the HBr generated by the allylic bromination is consumed by NBS to give succinimide.

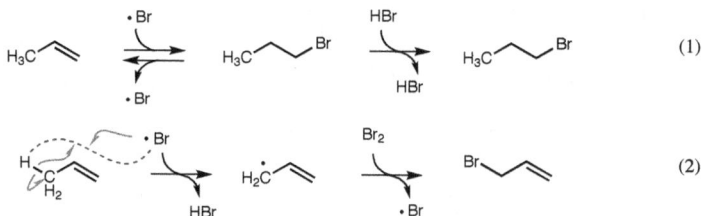

(1)

(2)

Benzylic Bromination with NBS

4. Benzylic bromination of ethylbenzene with NBS gives 1-bromo-1-phenylethane.

    Overall reaction

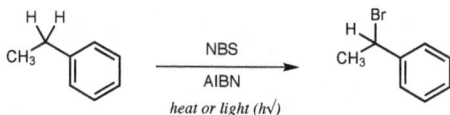

    Initiation with azobisisobutyronitrile (AIBN)

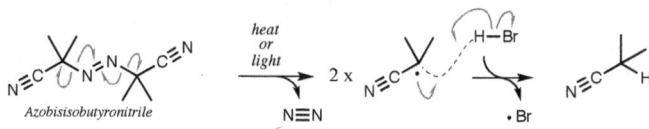

Propagation

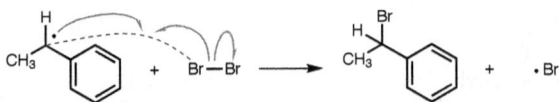

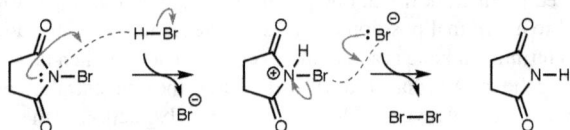

Termination

+ *others*

www.ingramcontent.com/pod-product-compliance
Lightning Source LLC
Chambersburg PA
CBHW072252270326
41930CB00010B/2358